Family Life in West Africa

50 Stories about Loving Africa

by Robin Edward Poulton writing as Robert Lacville

Published in 2024
ISBN: 9798337847993
Editor: Elisabeth Drumm (contact: drummediting@gmail.com)
Illustrations: Sarah Butts (contact: buttssarah5@gmail.com)

Copyright © Robin Edward Poulton

Robin Poulton owns the copyright for all writing by Robert Lacville. No part of this book may be reproduced in any form for commercial purposes without the written permission of the author: poultonrobin@gmail.com

Reproduction of any part of this book for educational purposes is freely permitted, on condition that the author and book reference are cited.

Other books on Africa by the same author:

Sunjata – Children of the Mali Empire ~ Then and Now (2020)
Twelve stories about real people that teach young and old about Africa, America and the cultural origins of African-Americans.

Sister Cities, A story of friendship from Virginia to Mali (2019)
Ségou and Richmond share 400 years of history, that has a new shape when Virginians visit Mali and Ségoviens discover Richmond.

Djita, a Malian girl from Virginia (2011)
This "flip book" in French & English, tells the true story of a young Malian-American girl. Aimed at elementary and middle school students, it is built around twelve exercises for teachers to use in teaching about the history of Africa and of African-American culture.

The Limits of Democracy and the Postcolonial Nation State: Mali's democratic experiment falters, while jihad and terrorism grow in the Sahara (2016)
This is a book about the political history of modern Mali since the year 2000.

More details about the author and his publications can be found
at https://robinpoulton.com/

The cover photo shows the author and his family in Timbuktu on January 1st, 1982.

The back cover photo shows the author's daughter in 1988, sitting with her best friend "Selina" and a bunch of the kids who all played together in and around our garden and swimming pool in Bamako.

Contents

INTRODUCTION 1

THE CHARMS OF TIMBUKTU IN THE SHADOW OF WAR ... 7

KILLING NOMADS, SAVING TUAREGS 13

WHO WILL HARVEST THIS RAINY SEASON? 18

GRILLING PRIVATE ENTERPRISE 23

IS BABY JEANNE A CHRISTIAN? (OR A MUSLIM?) 30

I'M SORRY ABOUT THE BABY AND THE MOTHER 34

TIME TO FEED THE CHILDREN 38

DIRTY WATER AND HYGIENIC WELLS 42

FIRE AND AIDS, SEX AND VENEREAL DISEASE 46

YOU WILL NOT EXCISE MY DAUGHTER 50

THE EYE OF YOUR NEIGHBOUR 55

WARM MONEY, COLD MONEY 60

MARITAL STORMS AND FORGIVING NAFO 65

SENDING NAFO'S KIDS TO SCHOOL 69

EDUCATION FOR EXILE..72

CHINESE MOSQUITO COILS AND POVERTY 78

A COY MOTHER TO BE 85

AN AFRICAN FAMILY BIRTH 90

WOULD YOU AGREE TO BECOME A SECOND WIFE? 94

MY DECEASED BROTHER'S CHILDREN'S SCHOOL..............98

BIRTHING AND CIRCUMCISING THE ROBERTS 102

EMERGENCY WARD TOURAY 106

IN THE RICE FIELDS OF TIMBUKTU 110

GOLD AND DROUGHT AND SEED BANKS.......................115

HOPELESS LOVE 120

HAPPY HUNGRY POLICE, A LETTER BY JEANNE LACVILLE
....124

A RESPECTABLE MARRIED WOMAN 127

RIVER BLINDNESS..130

SUGAR AND FRESH LEMON JUICE 134

DOGON ONION BALLS 139

DIAMONDS, BLOOD, AND CEMENT IN SIERRA LEONE .. 143

PEANUT BUTTER GRINDER 148

WHO OWNS OUR VILLAGE AFTER INDEPENDENCE? 152

APRIL FOOL 158

TRADITIONAL MIDWIVES 161

HOT RICE AND WELL WATER 164

CROCODILES IN THE BATHTUB 168

CROCODILES IN THE GARDEN 173

LIFE AND TIMES OF A GIANT TORTOISE 179

EUROPEANS SHOWING BAD MANNERS 183

IN AFRICA WE TAKE TIME TO SAY HELLO 188

ALBINO TWINS UNDER THE SUN 191

KEEP YOUR MAGIC HAIR ON! 195

RITUAL SACRIFICE IN MARRIAGE 201

AIDS, THE TWENTIETH CENTURY BLACK DEATH 205

DRACUNCULIASIS, A PLAGUE OF FIERY SERPENTS 211

ENERGY-WATER FOR GOOD HEALTH 216

TRYING TO FIND A YELLOW FEVER INJECTION 222

AFRICAN SOCIAL EFFICIENCY IS LINKED TO FRIENDSHIP 225

CHRISTMAS DRIVING HOT AND COLD 230

INTRODUCTION

During the years between 1990 and 2002, my column on *Developing Africa* appeared in the newspaper *The Guardian Weekly* under the pseudonym Robert Lacville - a frenchification of my name. The columns were popular, amusing, and filled with insights that were new to most of my weekly readership of around 350,000 people. I am grateful to the late John Perkin (GW editor from 1969-1993) for liking and publishing my Lacville articles each week, and even paying me for them. What a great hobby it was! While I was living in West Africa, I found new ideas to inspire me virtually every day. Africa is exciting, developing, constantly changing, always evolving and becoming less poor unless African leaders (currently in Ethiopia, Sudan and the Sahara Desert) destroy their continent with wars.

I wrote as "Robert Lacville" in order to separate my writing from my professional activities. Robin Edward Poulton worked for international organisations and governments. I was a senior technical specialist in various development fields, frequently meeting with high-level officials (including presidents, ministers, ambassadors, generals, rebel political leaders) to discuss delicate political subjects. Needing to be discreet, I could not to be seen writing articles in the international press. I did publish under my own name in *Le Monde Diplomatique*, arguing in favour of strengthening the social economy and working for peace with civil society organisations - but in a style and for a purpose very different to those of the Lacville stories. I also published political columns in a number of African newspapers. You can read about my professional career on my website robinpoulton.com.

For dramatic and creative purposes (and as camouflage), Lacville's wife Jeanne acquired a strong French accent and an imaginary third child called Jimmy, giving the vague impression that Lacville might be a Canadian. Otherwise, the articles were true to our everyday lives in Mali. The Robert Lacville series never described my work. I did not want anyone to confuse my professional life with journalism: I am not a journalist. I am a writer, commentator, and humourist inspired by the wonderful Guardian writers James Cameron and Ralph Whitlock. I describe what I see and analyse what is happening. A journalist reports and presents news; that is different from an opinion-based columnist looking at life.

The stories in this book all appeared in *The Guardian Weekly* during the 1990s. None has been updated to 2024: the stories describe the life we lived in Africa 30 years ago. They show our family life and my interpretations of Africa's development at that time. Many Africans have told me that they liked the Lacville stories and that my understanding of African life is truthful. When French scholars have suggested that I am an "Africaniste" I always say that "No, I am an African." I have not studied Africa in colonial university settings: I have become, through my youth lived in Nigeria and through my professional life in the lands of the Mali Empire, a White West African. I am proud of that. I love Africa. I admire African cultures and values. Mostly. No country is perfect and no person is perfect. As my friend Moussa Camara likes to say, *"Je me déplace avec les défauts de mes qualités"* = like everyone else in this world, "I have the defects of my qualities" - although I think it sounds better in French.

Western and African publishers are not easy partners, and twenty years ago I could not find anyone to publish the Lacville stories. Amazon in the 21st century has provided us with a new publishing outlet and I have been lucky to find in my friend Elisabeth Drumm a creative professional editor who is constantly encouraging and helpful, ready to help me create new books. She is rigorous in her edits and she also manages my blog. I am very grateful to her. Without Elisabeth, this book and several others would not have seen the light of day.

Dedication: to Ana and to Catherine Leïla

To my dynamic artist friend Ana Edwards, a brilliant Virginian educator and my partner in so many African and African American intellectual and educational enterprises in Virginia and West Africa;

      and also

To my fearless daughter Catherine Leïla Poulton whose career with UNICEF and with civil society organisations across the world makes us proud, mirroring the achievements of her mother.

Both Ana and Catherine Leïla are devoted to protecting girls and helping thousands of women through their selfless dedication and their amazing people-management skills. I admire them both.

THE CHARMS OF TIMBUKTU IN THE SHADOW OF WAR

The central mosque of Timbuktu was built in 1326. It is made entirely of mud and rocks. It is massive, and it is in perfect condition. Well, not quite perfect. After the good rains of 1991, the mud plastering on the outside is looking a bit shabby. Just now the mosque looks like the rest of Timbuktu: in decay. The city is full of abandoned houses. In the walls of the mosque, stones are starting to show through the mud plaster.... but not for long. January is the time for repairs and replastering. The great fourteenth century mosque is called *Jingeraiber* and everybody in Timbuktu is expected to contribute to repair and maintain *Jingeraiber*.

Once the Elders choose the day, every male must participate in the replastering of the mosque. It becomes a great social event. If any man does not take part, the others are allowed to come to his house and plaster him with mud as a punishment: social justice. Such brutal peer-pressure takes nothing away from the charm of Timbuktu. As dusk falls over the desert, the dunes glow golden around the city, the sun's last rays shine on the minaret of the mosque, and people lie down to sleep in the sand around the base of *Jingeraiber*'s massive fortress walls. Sand is soft, warm, and comfortable for those who have nowhere else to go. And the mosque's holy presence is reassuring. After 25 years of drought, thousands of nomads have lost all their animals, and therefore their livelihood. Destitute Tuaregs have been forced to come into Timbuktu. In this decaying city, there are a lot of people with nowhere to go.

I noticed during this latest visit that there are fewer people in Timbuktu than before. The market seems pretty empty and most of the shops are shut since they were sacked. Under pressure from the "rebels" returning from Gadafy's Chadian war, the army and the city's black population panicked against the Tuareg and Arab light-skinned traders (known insultingly as "red skins") and ransacked the market. Many traders were also money lenders. Years of usurious rates of interest combined with a new fear of the "rebels" have combined to ignite the powder-keg of history.

Since the fall of the Songhoy Empire in 1591 – when Moroccans with muskets defeated the Songhoy army at the Battle of Tondibi - the Black Songhoy (or Sonrai – local dialects along the Niger River Valley use different pronunciations) populations feel they have been subjected to more or less continuous red-skin domination and snobbery, not to mention the cattle-raids.
Then the White French came to impose colonialism, a new form of domination. Anti-French sentiment became anti-White—and with Independence in 1960 some of this Anti-White resentment morphed into a confusion between White French and Light-skinned Arabs and Tuaregs. History shows that the final anti-colonial resistance to the French was a battle in North Mali where the famous Chieftain Firhoun and many of his Ouillimeden Tuaregs were massacred in 1916. In fact, the French used Kounta Arab warriors (traditional rivals of the Ouillimeden) to kill the Tuaregs, while the French army arranged for it to happen. So that would be killing of White-on-White-on-White …. Complicated politics, conquest but not racism.

One thousand years ago, Arab raiders from Marrakesh captured slaves from all the clans around Timbuktu and Gao. Tuaregs, Songhoy, Fulani, Soninké and other language groups were victims of Moorish razzia tactics; it was the strength of the Mali Empire that allowed the peoples of the Sahel to resist these raids and establish 300 years of stability around the Niger River. Many Songhoy and Tuareg and Arab families are intermarried and intermingled, so the question of resistance to French conquest in only one factor in the social stresses in human relations. Occasional conflict is inevitable, though violence is not. A Fulani proverb expresses it best: "Your tongue and teeth are neighbours who work together every day. Despite their close complicity, occasionally the tongue is bitten by the teeth."

The conflicts in North Mali involve partly, but not entirely, the ancient story of friction between farmers who till the soil, and nomads whose cattle want to graze the farmers' crops. In fact, there are often age-old contracts in Mali that allow livestock access to stubble and river banks, and for cattle to spend the night on specific fields where their dung improves compost and crop yields. Some farmers pay herders to overnight on their fields. But conflicts emerge, and these days there are guns to add spice to disagreements. Herders and farmers no longer use sticks, spears and flintlocks: many have access to Kalashnikov automatic rifles.

The market of Timbuktu is not entirely empty. You can still buy cumin and aniseed, tea and sugar and tobacco, rice and wheat and millet, and rock salt carried in by camel from the desert salt mines of Taoudeni. Mutton or goat meat is available, although camel steaks have become rare because many herd-owning Tuaregs have left the city. You can still buy decorated leather shoes and bags, crafted copper tea-pots, delicately carved calebasses, swords and spears. Whole-sheep leather bags are available for nomadic travelers and great leather buckets sit in the sand waiting to serve customers who use 300-foot-deep wells (provided they have either a strong camel or two sturdy donkeys to lift the water from so far down).

The architecture of Timbuktu is charming, combining medieval Arab culture with Spanish-Moroccan styles from the seventeenth century. When Mansa Musa returned from Mecca in 1325-6, he brought with him a famous architect known as Al-Andalusi – at a time when much of Spain was under Moorish rule and Muslim culture. Al-Andalusi created the town house architecture characteristic of the city and also supervised the building of Jingueraiber.

In front of a typical stone-built Timbuktu town house with wooden studded doors and sculpted window frames I found a seller of charms. A "marabout," one of those local healers who mixes magic and religion with a knowledge of herbs and human psychology. I needed some souvenirs for friends and what better gist could I offer than a red pouch containing good luck and protection? Especially if the good luck only cost me a few dimes. I bought ten. For Ami's new baby I chose a ball of yellow cotton thread bound up with scarlet cotton thread, nice cheerful bright colors.... and it also has some pieces of chicken feathers poking out from one end. I saw another interesting shape. "What is this dark red leather waist band for?" I asked.

Most Sahelian women wear beads around their waist for luck and to excite their lovers. "This is a contraceptive" I was told. "The woman must wear it before the man comes to her and she must take it off before she goes to the washroom." Well, that is standard practice with these magico-religious charms. They must not come into contact with any polluting bodily excreta, so you must obviously remove them before washing. It is better to leave them in the bedroom since jinns and devils tend to frequent washing and excreting places: in the washroom they might get hold of your amulet, steal it away, and turn it against the intended human beneficiary.

Naturally I purchased a contraceptive belt. How could I resist it? Now I am in a quandary what to do with it. To whom can I give it? I certainly wouldn't risk putting it on my wife. Jeanne works in family planning and health. She'd be furious if I tried charlatan tricks on her. And I dread to think what would happen to our family planning if we put our faith in this loop of red leather containing a few pieces of magico-religious paper. Nor would it be kind to give it to a friend, especially if they took its supposed powers seriously. I shall keep the contraceptive belt as an anthropological curiosity. But I did get another product that might come in useful for a friend. When I asked the charlatan for something to make babies, he sold me a dark green powder: one pinch to be taken in porridge (by the woman) before the procreational act. I have tasted it: it is gritty and dusty and tastes like pounded bush leaves. It is intended for women. Do you think anything will happen to me? If so, what?

Not all of this magic medicine is bad. I reason that the charms must be pretty good in the region of Timbuktu since the medical services are practically non-existent. Since the people of Timbuktu have had to live for the past 1000 years without any functioning medical clinic, what else could they rely on but magic? Apart from some foreign charities, neither donors nor government have invested in the region's development since Independence more than thirty years ago.

The African Development Bank has loaned money to the government for a new regional hospital building in Timbuktu, and this building has just been inaugurated. However, the hospital has no doctors. The ADB has simply added to its depressing record of creating debt-without-development. Not for nothing does the ADB's French acronym come out as BAD – as several educated citizens of Timbuktu told me.

"Why is a regional hospital so useless?" I asked these citizens and officials who were fuming against it (including some of the most senior government officials in Timbuktu). "To start with, it has no beds, no equipment, no nurses, no doctors," said an old friend with whom I used to work in this part of the world. We were taking dinner under the stars, enjoying the cool desert air on the flat roof of his elegant Timbuktu town house. Modou stretched out on his mattress, adjusted himself against his leather cushions, and sipped his mint tea. "Secondly it cannot serve as a regional hospital since there are no roads and no means of transport for sick people, except donkeys. Therefore, it is only a hospital for the city of Timbuktu. But we have just built a beautiful new city clinic which does have a doctor: so why in the name of Allah would we need a second one? No, my brother, this magnificent so-called hospital is what they call in English "a red-skinned elephant."

So, who benefits from the new hospital? Modou explained that the ADB (like the World Bank) works primarily to benefit Western corporations. The contract was awarded to a French construction company that takes 25% in profits. The French would have paid 10% to the former Minister of Health who pushed through this project without consulting anyone in Timbuktu and some of this "commission" would filter through the military junta and the sticky fingers of Mariam Sissoko, the wife of general-President Moussa Traoré. Modou works within Moussa's One Party State, so he knows how the monies flow.

But Modou is also keen to support his own people. "If they had given us the millions instead to the French, we could have supplied local health services throughout the region for several years instead of building an unnecessary hospital building." No real surprises. Just another donor project funded by the plump to fatten the corpulent while loading the African peasant with more debts.

I looked over the range of amulets on the woven mat beside the Friday Mosque and chatted to the charlatan. Like most northern marabouts, he was called Cissé. He offered charms for the young and the old, charms for men and for women. I listened to other people buying. There was one for back pains. I said I might buy it. Cissé assured me that in addition to curing back pains, it also would do wonders for my virility. I couldn't get my money out quickly enough. Africa's great phobia is impotence. I have bought myself a guarantee for only five hundred francs cfa (one pound sterling or around 50 US cents). And no more back-ache into the bargain! In Timbuktu the Mysterious, you can combine health and virility for less than two dollars. So, the local health services are not so poor after all. Would any of my friends like one of these miraculous belts as a new year gift?

KILLING NOMADS, SAVING TUAREGS

The Sahelian drought cycle started in 1965. There have been two great droughts, in 1972/73, and again in 1983/84. The greatest victims were Tuareg desert nomads. Sure, the drought years were tough for everybody. Sedentary farmers went hungry; their children died; their sons were forced to seek money for the family by working in the urban jungles of Abidjan, Accra, Lagos .... some even went to find work as far away as Libya or the Arab Gulf States. But now in the 1990s, the rains are respectable again. Families are back growing millet, sorghum and maize in the fields of their village.

But many nomads have no fields. Their wealth is their livestock. When the rain fails, the grass fails. Wells run dry. When there is no grass and no water, the sheep and goats and donkeys die. During the drought, the animals perished. In the late seventies civil society and charitable organisations helped nomad families to rebuild their herds, and quite a lot of nomads have managed to return to their life of herding in a small way. Then came the drought of 1983, which provided the coup de grâce to the nomad lifestyle. Some Tuareg groups believed the end of the world had come and took off for Mecca, where a pious Moslem hopes to die. Most of the rest became refugees on the town fringes along the Sahel, waiting for the World Food Program of the United Nations to give them a starvation ration.

When European and American food aid arrived in the 1970s, it was stolen by corrupt officials and colonels running Mali's military dictatorship. In 1968, the independence leader and politician Modibo Keita was overthrown in a coup d'état led by an army officer called Moussa Traoré. His colonels sold the food aid in neighbouring countries, and built houses in Bamako with the proceeds. Without the food, many Tuareg women and children died of starvation along the Algerian frontier. They have never forgiven Moussa Traoré or his army.
If there has been trouble through 1990 in the northern regions of Niger and Mali, in the south of Algeria and Libya, it is because

there are thousands of destitute Tuaregs with nothing to do. North Mali is close to Algeria, where Tuaregs from Mali (and from neighbouring Niger) arrived as impoverished refugees. Some young Tuareg men travelled further north into Libya, where the dictator Colonel Muammar Gadafy said they would be welcome. Gadafy armed some of them and trained them as soldiers. In 1990, some of these soldiers came home to make trouble. "We were born in Mali," they say: "We have come back to claim our birthright, and to take revenge against the military dictatorship that left our mothers and sisters starve to death."

There are few things sadder than a destitute family of refugees. Nothing is more pathetic than a proud people diminished. Where should desert nomads go, if they have no animals to nomadize with? What is to become of the Tuareg "blue men of the desert" with their indigo turbans, their sculpted camel saddles, their dark desert tents ablaze with wonderful leather decorations? This is the story of a few of them, nomads who have found a new life as farmers. It is also the story of two pioneers, a teacher called Mohammed Ag Attaher and a young British volunteer called Bernadette.

If you drive along the new tarmac road northward out of the riverine trading city of Mopti, you can see through the distant heat haze a water tower situated beside the city refuse dump. This is where a group of Tuareg refugees were living. They were scratching a living in the refuse heap, while waiting for the next United Nations food hand-out. Mohammed Ag Attaher is a Tuareg from Timbuktu. He happens to be a primary school teacher in Fatoma village beside the tarmac road. There is some spare land along that road and he reasoned that his fellow-nomads would be better off living on a patch of decent land, rather than on top of the rubbish tip. So, Mohammed cajoled and persuaded and encouraged a few families to leave the water tower, and to follow him to Fatoma village. Now there are fifty families installed there. Mohammed is persuasive. He obtained help from half a dozen charities, from the Catholic church, but mostly from his own people: and six years on, there is now a new village where previously there was only sand. People are building houses. Mohammed himself occupies the teacher's house in Fatoma; but he has also built a house in the new village. Mohammed keeps one wife in each house: it is better to keep co-wives separate, to maintain harmony.

Since arriving in Fatoma, the nomads have planted 11,000 trees where there were none before: and 9,000 seedlings have survived. The nomads have learned to farm. Their millet crop was fair this year, although a plague of grasshoppers destroyed their rice crop before harvest. When I visited Fatoma before the rains began, a grant from the European Commission was being used to finish two deep wells that will make the villagers self-sufficient in domestic water.

A few miles away, close to the Niger River, another nomad village has sprung up in the midst of nowhere. A small Malian voluntary organisation called OMAES has created a resettlement program with the help of Christian Aid, Bandaid, and Bernadette, a volunteer with the United Nations Association of the UK. I first came here in 1987. There were fifteen miserable straw shelters and a lonely British volunteer in a pickup truck struggling against the weather (no rain, nothing but the burning destructive sun), the insects (locusts that year, and mosquitoes by the million), the bureaucracy (no one would give anything to the nomads: it was always another department's responsibility), corruption (the Administration of Mopti region would only allocate land, if Bernadette would pay them 12 000 dollars to "survey" the area), and the donors: yes the generous donors are always part of the problem. In this case, it is because refugees are only "sexy" when they are a new problem. Refugees become a nuisance when they are no longer a hot topic on the BBC News. And the grasshopper invasion did not excite anyone: a plague of locusts is sexy, so grant money to fight locusts can be obtained from FAO. Grasshoppers are not as sexy, even though they eat just as much as locusts!

Another problem: right now, Africa has lost its sex appeal for Western Governments. In the 1990s, since the fall of the Berlin Wall, Eastern and Central Europe are walking away with the Miss World development prizes, while the destitute nomads of the Sahel are being left to starve.

But Bernadette refused to give up. Little by little she led the nomads out of destitution. By careful economy, she found the money to recruit a small team of OMAES workers. With the help of Salif, a smiling black agronomist from the south, and Christophe, an optimistic pale young Tuareg from the north, she was able to provide the Tuareg nomads with training in farming. Wells were dug and a few women began to grow vegetables, hand-irrigated with water from the wells. The ex-nomads chose a bright young man from their group for health training, and he began to provide first aid and hygiene instruction for mothers. In 1987 when I first met him, he occupied the best of the straw shelters and he was proud of his tiny stock of medicines lined up on a home-made shelf of sticks and string. In 1990 he occupies a mud-brick house, and his medicines are carefully stored in a wooden cupboard. Progress!

But it is the spirit of the community that has changed most of all. In the beginning, when Bernadette rescued these nomad refugees from their refuse tip beneath the water tower, she was leading a ragged bunch of beggars. She has now returned home to Britain. But when I went recently to visit the village with Salif and Christophe, I found a cohesive village community. And I told them: "When I came here to meet you back in 1987, you had only one kind of talk: you said 'Give us help, we are poor people, we have lost everything, we need your help.' But today you have a different kind of talk. Today you are saying 'We need a higher dosage rate of fertilizer for our millet crop; we need to be more careful to protect our tree seedlings from the goats; we must accelerate the building program.' When I came in 1987, you were refugees and beggars. Now in 1990 you have become farmers and villagers."

In every village, the spokesman is an Elder. Here it was Sayeed who answered me, his eyes bright in his weathered face, beneath thin grey hair and an ancient embroidered skull-cap: "Yes, it is because Bernadette showed us. She is our mother."

Christophe said: "It is true that we have done good work here. The people have enough to eat, and they have recovered their self-respect. But we must remember that they have been forced to change their whole life. They used to be free nomads, ranging the desert with their flocks. Now they are tied to the land, to their crops, and to their debts. We have been saving their self-respect, but our project is also killing nomads."

WHO WILL HARVEST THIS RAINY SEASON?

The thunder roars, the rain pours, Africa changes. It is May 1991, as mango trees go green and the red dust of Africa washes off the leaves. When the wind rises before the rains, our cat begins a frenetic panic run into the garden. Once the rain pours down from the sky, our cat reckons the house is a safer place to be. At the door, she stops to hiss at her enemy the dog. The dog is crouching terrified inside the house: she is allowed inside only on days of thunder. The cat spits viciously, then races through the house before the dog can react. But the cat is making all this fuss for nothing: Sac-à-Puces disdains to take any notice of a mere cat. She is an old dog now and in Africa age carries status. Age brings wisdom, inviting serenity and respect.

I suppose that, out of respect, we ought to change the dog's name. "Mother", or "Old Woman" are honorific titles which we offer as compliments to elderly ladies; maybe the dog has earned such an honour too. Her current French name of "Flea-bag" was given her by some disrespectful British volunteers twelve years ago or more. In her youth, it was amusing. Such a flippant name cannot be easy to bear in your old age. One American Peace Corps volunteer, believing that her French was improving thought her name meant "bag-of-pus." Now that really is insulting. Who would ever call a dog a "pustule," a sack of liquid putrefaction? These Americans... really.

As a matter of French precision, if she was really a "pus-bag," the French would be "Sac-a-pus" pronounced "pue" (which is almost like "pew"). If you write it "pue," it means "stinks" in French. Our dog would become "Stink Bag" – as a name? Oh no! Please! Fleas are bad enough. When these anglophone idiots come around and mispronounce her name, it becomes shameful!

"Flea-bag" is already a tough call. Our poor dog has not carried an easy burden these past twelve years. Yet to carry such a burden brings serenity in later life, experience that allows you to disdain a mere cat.

Our cat is rather neurotic. When the wind starts to blow up before a storm, and the slender eucalyptus trees bend double with a deafening rustle of leaves and falling branches, a black streak flashes through the doorway, zig-zags frantically between the falling leaves, flies over the low wall beside the jasmine bush, and vanishes beneath the passion-flower creeper. Then the clouds are gathering black and mauve in the Eastern sky, and nature's disco-show begins. Flashes of light illuminate the massed rain clouds. Streaks of forked lightning split the sky. The lightning emblazons the dark blue-mauve-grey sky like silver embroidery on a giant West African festal robe. My daughter cowers against her mother while the three-year-old hides under the blankets. But I love to stand on my thatched roof terrace, keeping dry as I watch Africa clothed in its most magnificent attire. When the wind blows up to announce an African rainstorm, I go out to admire the Greatest Show on Earth while my cat takes off in blind panic.

The dog takes a different view. Condemned to live in the garden, she lies for long hours in the deep shade under the purple bougainvillea, conveniently located near the cool damp of the outside garden tap. Many village girls and women come to fetch water at this tap, so there is always water and mud around there. Evaporation makes this a cool spot. 120 degrees F in the shade does not suit dogs... nor flea-bags, nor humans, nor any creature I know except Margouillat the orange-headed lizard whose head bobs up and down on the top of the wall, looking for unwary insects and watching in case the cat is around. Margouillat needs the sun. In the morning, he is too sluggish to be safe and has to take his chance on the wall until the sun has warmed his blood and set his agama muscles into their normal lightning agility.
When a storm threatens, Sac-à-Puces knows that she is allowed inside the house. She approaches the door looking humble, head down against the dust and wind. She rolls her eyeballs up

at the humans who are carrying chairs and mattresses into the house, and bringing dry washing out of the impending storm. When the thunder rolls, the cat goes mad in the garden while the dog creeps inside and curls up at the foot of our carved statue of an African hunter.

Then the first drops fall. They are big floppy drops, warm and wet. This is no pleonasm. The rain in Africa really is wet. Tropical rain is much wetter than northern rain. European drizzle is cold and mean and insidious. It gets into the joints. First you feel cold. Then you realize you are damp— your protective clothing has been letting in drops at the feet and neck, your clothing has become wet and you have probably caught a chill.

African rain is warm and friendly. Here in Mali, we have honest rain, frankly big and wet. West African rain doesn't make you cold. It leaves your joints intact. The rain reminds you that you were missing rain's friendly presence; reminds you that you are thirsty, that there has been no rain for eight months and you are glad to welcome back warm and friendly rain. After eight long months of heat and dryness, the earth breathes, the air cools, the grasses begin to sprout.

Farmers have already cleared the weeds and brush from their fields. They have already spent long afternoons in the heat of April and May, digging up grasshopper and locust eggs and burning them. Or else they bring the eggs to the Catholic Mission, which is offering one bag of fertilizer for each bag of grasshopper eggs as part of the national campaign to save the harvest from the predators. Now the rains have come to soften the parched soil. After the second rain, the seeds are taken out of the communal seed store and oxen are harnessed, and the long trudge of plowing begins. Those who cannot afford a plow (at $100 for a decent quality steel plow, or $40 for a poor-quality local blacksmith's product made out of recycled car doors) are forced to dig by hand. The back-breaking work of hand-hoeing has been the basis of Africa's agriculture for millennia. Animal traction makes a better and easier job (while tractors are a menace, destroying the fragile soil and promoting erosion).

The rains also usher in the "hungry season." This is the time when the farmers need most energy for their hard labors: yet in many regions the grain stores are empty after the poor harvest of 1990. Food distribution has been carried out into many deficit regions. But in the north around Timbuktu, the Polisario Front and Col. Gadafy have sabotaged food supplies by sending in armed raiding parties. This is more than the traditional annual "cattle-stealing razzias" by Moors and Touaregs. This year there is serious shooting of soldiers and civilians. This is no cattle business: this is Kalashnikov business. This is not the raiding of a traditional "hungry season:" this is a "political hungry season" where hungry political figures are looking for fat pickings.

The American charity CARE says that the first cases of child starvation around Timbuktu already started in April, but they are described officially as "illness." If you start hearing about "starvation" instead of "illness," that will mean that the region of Timbuktu is succumbing to a general state of war, and Gadafy's planned sabotage of the Sahel will be going nicely, thank you, as planned.

This means the Sahrawi referendum will be under threat: Morocco claims former Spanish Western as its own province, while Algeria (partly to annoy Morocco) supports the idea of an independent Sahrawi Republic. If this breaks out into war, Mali and Niger may come apart at the seams. Chad has had periodic civil wars. Gadafy says all Tuaregs are welcome in Libya. He is encouraging racial violence will throughout the region. We have recently seen friction between Mauritanian and Senegal, partly due to problems involving nomadic herds and farmers' field crops. National solidarity is fragile.

Gadafy is trying to change the political map of West Africa. The political harvests may be rich for the ruthless. If armed violence breaks out, people will be forced to migrate south. Farmers will not be able to weed their fields and harvest, even if their crops are looking good with the grasshoppers are under control. In that case the 1991 food harvest will be a failure, too, and all that lovely rain will have been wasted.

GRILLING PRIVATE ENTERPRISE

I have been helping a Malian friend to set up an enterprise. Quite a small enterprise, you understand: in the catering business. He employs three or four people, with extras from time to time. He is an expert in grills. No one does a grilling like Seydou does a grilling. And if more is needed, such as a juicy onion-and-lemon chicken yassa, his wife helps with the cooking. Africa needs entrepreneurs and there are not that many of them. Local culture does not promote private initiatives: traditional education provides answers that the Elders have devised over centuries of experience and it does not encourage creative thinking.

Individuality does not exist in the same way that Americans conceive of personal identity. A West African is born into a family unit and she or he has no real identity in traditional African society outside that family unit. Boys and girls receive at birth a place in the family hierarchy, a status conferred by their parents, and a set of instructions (called "education") to keep them where they belong. Seydou was also born into a caste: he is a blacksmith, and he knows his place within the hierarchy. The Elders and people with wealth and high status govern society. Young people are expected to know their place and to keep silent. Individualism and initiative are not encouraged.

Names are a good clue to this state of affairs. In the village, you will seldom hear a mature woman called by her given name. Among the women, they are usually call each other "the mother of Mohammed" or "the mother of Fanta." The original Lion King, Sunjata Keita who founded the Mali Empire, was really named Prince Djata. His mother was named Sogolon, and so as a child he was called Sogolon's Djata—which fused into "Sonjata" or Sunjata.

Among the men of a village, a woman will be known as "Amadou's wife." In the family unit, men call their wives "wife," wives call their men "husband," or "old man." It is important to remember that "old" is a term of compliment and respect in Africa, not a synonym for "senile" or "useless" or "dodderer": we in the West have forgotten so much that we could relearn from African society. On the other hand, they could learn something from White societies about promoting initiative.

Even when one of my friends uses his given name, one has to ask the question: "Whose name is it really?" Is it really the name of the child? I am very fond of an elderly lady called Djita (although I would never dream of calling her Djita; I call her "Mother") who has two very beautiful daughters. But conversations are sometimes confusing because Djita refers to them as "My mother" and "My mother-in-law." Each daughter is named after a senior member of the family: and because the daughter carries the name of the grandmother, Djita calls her "mother"! Here is a typical piece of Djita's conversation: "I have decided to let my mother-in-law sell these cloths since she has a better head for commerce than my mother." This really means that one daughter has diplomas while the other has sales technique.

Then her husband comes home for lunch. I call him "Papa" and we will share a plate of rice and sauce. Papa uses the same system to refer to his daughters. Except that, for him, the "mother" and "mother-in-law" are the other way around. It is a real head-twister!

In Cameroon they do not even ask the question "What is your name?" They ask the question: "Who are you?" You may reply "I am Jim".... but you are just as likely to reply "I am Grandfather," or "I am Uncle Jim-Biya." And if that is your reply, then who are you in fact? Are you you? Or are you Uncle Biya? Since Uncle Jim-Biya is dead, and since the ancient African cosmogony is based partly on the idea of spirits being reborn.......then you may well be Uncle Biya. In any case, if you have been called Uncle Biya since your birth, and if you have been brought up to believe you are Uncle Biya, then you probably are Uncle Biya. Like the Dalai Lama of Tibet, you are

carrying a pre-determined identity and therefore a destiny. Whatever the family expects Uncle Biya to do, that is your destiny. You will carry seven of the characteristic qualities of the deceased namesake and they will have been drilled into you from the earliest days. Under such circumstances, it is not surprising that private enterprise initiative is a rare commodity in West Africa.
[Unless, by chance, your Uncle Jim-Biya was an entrepreneur.]

So, when I meet a young entrepreneur in West Africa, I like to try and encourage him. Or her: no one can accuse me of sexist bias in this field. I know all about "gender considerations," thank you! I am actually in partnership with one supposedly entrepreneurial lady, trying to help her launch her commercial career. It has cost me a pretty packet so far and we seem to be getting nowhere. At least, not commercially. If her business ever breaks even, I'll write the story of her/our success. But this story is not about my business woman troubles, only about my catering friend Seydou. In this case, Seydou has experience. He has worked in hotels the length and breadth of West Africa and the Costa Brava, he has Major Domo'd for Ambassadors and Ministers, and now he has started a restaurant and catering business.

Seydou has lots of good ideas. He has a small fixed restaurant. He does catering for parties (I have used him for a couple of parties myself) and for official receptions (where I have been helpful by introducing him to White Embassies and American Project Managers); he feeds seminars as well (there are lots of seminars in Africa: there are nearly as many seminars as Project Managers). But I would hate to give the impression that Seydou depends on me. He created this business himself and he finds the contracts himself.

The time-honoured system for getting contracts is through connections. Why else do architects and building contractors stand for election to local councils in Europe and America? In Africa, connections are based on the village and the ethnic group. Seydou happens to be a Fula. The director general of the Central Bank happens also to be a Fula and hails from the same northern desert town as Seydou. What could be more natural than that Seydou should call on him to talk about his New Year party? The director general called his administrator and suggested that Seydou be given half the catering contract this year. Naturally the administrator agreed with his boss. But he wasn't very pleased since the previous year the whole contract had gone to some ladies from his own village in the south. To cut a long story short (and believe me, it was a long story: when Seydou told it to me the other evening it took him forty minutes to recount every blow and insult), the administrator tried to cut Seydou out, Seydou fought back, and the director general gave Seydou the whole contract. So Seydou had to purchase, slaughter, prepare, and grill 1200 chickens and 60 sheep.

"You mean that guests to the Bank's New Year party consumed 1200 chickens and 60 sheep?" I asked incredulously. "Even if they had 400 guests, which is unlikely, that means that each guest would scoff three whole chickens, and that's without counting the sheep!"

"No, no, no, you do not understand. Only 100 chickens are for the party, and 10 sheep. The rest were presented in nice cardboard boxes for delivery to important people. After all, they have their families to feed. And some of them are unable to attend the reception. And if they come to the reception, it is unseemly for Big Men to fill their pockets with food for their families. So, we send round the food for their families and they can enjoy the reception even if they do not attend. The director general was very pleased with my performance. I am sure I shall receive the contract again next year."

I think back to my commercial days. Fond memories of living well on the private sector before I went off with Jeanne into the African bush. Good customers used to receive courtesy boxes of booze for Christmas, so did the managers, and so did good salesmen. Once we took the sales force off to Spain for training. Well, let's say for "training." If it was straight training, any local hotel would have been fine. This was training + motivation. Spain has kudos and sex appeal. Hormones don't run at home like they run in Spain. Our "training" session had more than a smattering of other ingredients like "motivation," "prestige," "reward," "bribery": things like surfing and sun, wine and a week-away-from-the-wife (only men on this junket), and free tickets for the sales managers ...... It is reassuring to sense that the Central Bank is moving towards the private sector. Perhaps 1200 chickens is a little meagre. 60 sheep seem rather inexpensive when I compare the cost of flying our sales force across Europe to Spain. Of course, it was all written off as a business expense so the tax payer funded most of our junket in Spain: it was really a form of legally-approved corruption. And, of course, tax payers covered the cost of the 1200 chickens and the 60 sheep. Well done Seydou for landing the chicken contract. Perhaps it will lead to bigger things. If the Central Bank goes private, the managing director might fly their staff to Spain or the Cape Verde Islands.

Contrast this happy affair with Seydou's even bigger contract with the public sector Shipping Company. Seydou had the idea that the Timbuktu cruise steamers could do with improved catering. He is dead right. I can still smell the lunchtime uric acid wafting heavily into the ship's dining room from the second-class toilets, strategically placed next to the kitchens where the greasy rice is served daily, washed down with tepid water. I found it necessary to pinch my nostrils while eating and drinking or would never have got it down, never mind keeping it down. The ferry to Timbuktu and Gao is a memorable experience. I recommend taking your own picnic. Alternatively, if you descend to the bowels of the ship where the fourth-class passengers travel, you will find women cooking tasty rice and sauce dishes (bring your own dish) on live braziers, literally at the water line.
It is also the fourth-class toilet facilities that I use because the second-class porcelain installations upstairs beside the dining

room are simply too disgusting to contemplate. And third-class is smellier still. However, the fourth-class passengers (plus me) have a cleaner and less smelly defecating system: there is a board and a rope with the surface of the river splashing just below your naked backside. Your feces drop straight into the Niger River water to feed the catfish and the Nile perch. Just don't let go of the rope!

These were the ships that Seydou hoped to feed. He went to the shipping people to negotiate his contract. Cabin passengers purchase a full-board fare, so Seydou would receive money for his restaurant services from the Shipping Company. The contract was signed and Seydou went off to invest his savings in food and drink. Three months later he turned up in my house looking haggard. The Shipping Company now owed him $9000. Seydou had no more savings and his credit was exhausted with his suppliers. The Shipping Company is owned by the Malian State. Money earned by the company goes into the Ministry of Finance. Very little ever comes out of the Ministry of Finance. Seydou is resigned to waiting months, perhaps years to get his money. Could I advance him 10 000 Fcfa ($50) to feed the family this month? Banks and private sector people pay their bills. In the public sector, they don't.

The dominance of the public sector in most African economies bogs down private enterprise. A few countries like Kenya and Zimbabwe, Nigeria and Cameroon have private sectors that now provide the dynamic for the national economy. Most other countries are either too small, too poor, or too badly managed (or all three) to have developed a decent private sector. In addition to that, Mali is too French: the French bureaucracy is famously cumbersome and opaque; general-President Moussa Traoré's military One Party State has perfected the opaqueness.

Quite a number of African countries have an expanding social economy (based on cooperatives, village associations, charities and voluntary organisations) which offers hope, but too often we find that private sector initiatives are squeezed out by the State.

"If they see you are having some small success," said one of my business lady friends who runs a hair-dressing salon in Bamako, "the government officials come down and collect all your profits in bribes and taxes. With the IMF economic squeeze, no one is spending money any more. Women are buying wigs, which we dress at 1000 francs a time ($4) instead of their dressing own hair, which costs them 7000 francs. Thank goodness it is my husband who provides our main income! Last year I just about covered my rent and the salaries of my three employees. Then the State came and asked me to pay taxes bigger than my turnover. If they insist, I shall just have to close the shop and my three employees will be on the street."

Seydou's little catering business is threatened too: not by State taxes, but by State debts. In Africa you negotiate through parental links or village affinity. With the Central Bank, Seydou can negotiate with his fellow townsman. They can discuss in Fulfulde, their common Fula language. Payment is true and quick. But with the Shipping Company Seydou has no recourse. I suppose he could go to court, providing he has $1000 to spend and can afford to wait two years for the case to be called. But that would be throwing good money after bad. Too many private sector initiatives have foundered on that rocky salvation. No, Seydou has to badger them, beg them, flatter them, or find someone in the Ministry of Finance who will support his claims. I reckon the public sector Shipping Company has grilled poor Seydou, burned him up good and proper.

IS BABY JEANNE A CHRISTIAN? (OR A MUSLIM?)

One week after Awa gave birth, sitting on the floor in her house attached to our house, the baby was named. Her father Nafo decided on Jeanne. No great surprise there: my wife is called Jeanne and since she is the family benefactor, the most natural thing in the world is that the little girl should become Baby Jeanne. Her brother is already Ousman Robin. Our obligations to the children could not be clearer!

But how should Baby Jeanne receive her name? Nafo is neither Christian nor Muslim. The "Ousman" part of his son's name shows clear Islamic influence, but it was chosen by Nafo's big brother who is head of the family and an important man with four wives. Nafo did not really like the Islamic touch. He is Senoufo, one of the Malian ethnic groups that is most attached to its community traditions, to its Poro initiation system, to its history. Nafo's father was Head of the Poro Society in his village, near Sikasso. That is a very prestigious position. Why would Nafo be interested in imported colonial religions?

Islam came to the Sahel long before Christianity. It seeped in through trade with the Maghreb. As early as the 9th century AD (2nd century after Mohammed The Prophet), Arab camel trains brought salt and cloth and Islam to trade for the gold and the slaves of the Ghana Empire. By the year 1200 there were many Fulani, Haoussa, and Manding converts living in the Kingdom of Soso and then in the Mali Empire founded in 1235 by the Lion King – Sunjata Keita. Mansa Musa, the most famous of Sunjata's successors (and probably his great-nephew) made a pilgrimage to Mecca in the year 1334 which brought Mali to the attention of the Mediterranean world. Thanks to his control of Mali's gold mines, Mansa Musa has been calculated to have been the richest man the world has ever seen: far richer than Bill Gates or Elon Musk, richer even than the two of them together. Mansa Musa distributed so much gold in Cairo and

Mecca that the price of gold in the Middle East did not recover for twenty years!

The first serious arrival of Christianity in Africa dates only from the baptism of the king of Congo by the Portuguese in 1484. Islam and Christianity are not accepted everywhere: the Senoufo (Ivory Coast & Mali) and other groups resisted conversion; the Bambara (Mali) were converted to Islam by force in the 1850s; the Wolofs (Senegal) accepted Islam even later, under French colonial occupation.

Senegalese Islam is particularly interesting. Islam is split down the middle, between the Tijani Brotherhood and the Mourid Brotherhood. The latter seem to be gaining ground all the time. Senegal is the scene for a religious struggle to win the hearts and the economic power of the people. They say the President of Senegal consults with the big religious leaders more often than with his Ministers, especially with the Mourids, who control the country's main export crop (groundnut) and an ever-widening range of other economic activities.

Mourid is presented as an African Islamic movement resisting "Arab imperialism." This was spelt out for me by my Mourid guide in clear syllables as I climbed the minaret of the great mosque in Touba, the capital of Mouridism. The whole religious complex (college, library, mosque, courtyards, housing) was built with Senegalese money, contributed by the faithful. Never have I seen a cleaner place. Nowhere in Africa is a city better organised. The delicious irony of Mouridism is that its founder Amadou Bamba created his religious brotherhood in a spirit of resistance against French colonialism. Now the Mourids have become a movement "against Arab religious imperialism," to quote my good friend Lamin. Most of Africa's Islam is dominated by Saudi Arabian money. Bamako's old Friday mosque was torn down in the 1970s and replaced with a Saudi-funded mosque (which made a lot of Malians very angry). For the Tijiani Brotherhood in Senegal, the Saudis built - in the city of Tivouane - the largest concrete mosque in Africa. In reaction against this Arab religion of the cheque book, Amadou Bamba's grandsons have maintained their ancestor's spirit of rebellion, by creating an African religion of protest against imperialism. The Mourid message contains inherent elements of independence and protest.

As the Mourids are winning in Senegal, so Islam is winning in West Africa. The reasons are many. First of the reasons is the Islamic tolerance of polygamy. Africans have always been allowed several wives: the obligation to take over the wife of your dead brother (technically known as "levirate") is a major part of the African safety net for widows and orphans. Christian strictures are more difficult to live with than the Islamic limit of four wives. Islam is also strong on brotherhood, which is very African. Any Muslim can arrive at a strange mosque, sleep there, receive food, and pray shoulder to shoulder with his hosts. In how many Christian churches would one receive such hospitality? The Islamic communal prayer is also very strong. Kneeling shoulder to shoulder together in the mosque, you cannot fail to feel that sense of community, which is both Muslim and African. Finally, Islam has had the advantage of appearing as the anti-colonial religion (as compared to Christianity, the religion of European colonisation). All these factors explain the progress of Islam.

It was in the middle of Muslim Senegambia that Baby Jeanne came into the world. Nafo and Awa were far from their home in Mali and far from their families. Their Malian friends were numerous, and Muslim. The men gathered each day around Nafo's cigarette and sweet-selling table outside our house. The women lay in the shade of the giant mango tree where Awa plaited their hair into wonderful shapes. After the birth, however, this concourse paused. As Awa stayed inside to suckle her baby for a week, we wondered how Nafo would decide to name the baby. Would he have a ceremony? Would it be a traditional Senoufo ceremony? Would he kill a cock? Would he conform and have a Muslim ceremony? Would he come and ask us for the money to slaughter a sheep? On the 6th day, he told us to prepare for the morrow, at 9 a.m.

We arose at 6.30, and emerged into the garden at 7 a.m. after breakfast. Nafo told us that Baby Jeanne had been named. When? "Oh, I ask the Imam to come and he say 9 o'clock, but he come 6 o'clock." The imam is an elder, the senior dignitary of the village. It is always the older men who lead ceremonies. The imam is also a Muslim religious official, head of the local mosque. Had Baby Jeanne been christened a Muslim? Nafo shrugged with indifference: "Islam no good for me: but family too far away. My friends say better get Imam."

So Nafo had bowed to social pressure, like so many parents who are nominally in the Christian Church, and who only turn up for christenings and funerals. He had to do something .... so, he turned to the dominant local religion. Had he been in Zaire, he would have called the priest. Here it was the imam, acting as traditional village elder. Thus, Baby Jeanne received her name before breakfast.

Later when we returned to Mali, we discovered that the baby had received "Aminata" as her second name. How and when she got her second name, and from whom, was never fully explained. It is probably the name of a grandmother. In any case, the name Aminata - with its Muslim resonance – allows her more flexibility, which will be a good thing for her in a West African world that is becoming increasingly Muslim.

I'M SORRY ABOUT THE BABY AND THE MOTHER

Diallo came to say he couldn't make our afternoon rendez-vous with the children. "I have to go to a funeral," he explained. Well, that might delay us a bit: but funerals in West Africa are very common, for you always show solidarity with happiness and with sadness. We all attend at least one ceremony per month to support a neighbour, colleague, friend, or one of a friend's distant cousin's in-laws.

"But this funeral will have to take me the whole day, says Diallo. You see, it is for the infant of my brother. My own brother, you understand, same father, same mother. The infant died this morning in our house." I express sadness and shock. Diallo thanks me, and continues: "It is hard for him, for all of us, because it is his first child, and from his second wife. The first wife was unable to conceive, so my brother took a second wife. Now the second wife gave birth yesterday in the afternoon. I was happy for my brother, and I immediately sent word to all my family and friends and neighbours, inviting them to the baptism in one week's time. Then this morning my brother called me in my bed. He said to come and see the child. It was dead."

Poor Diallo. It is he who has a son with emotional troubles: the child was rejected by his mother, Diallo's wife. Now this! Alas, he is just one of a million families with the same problem. What could be worse in the world, than to lose a child? A fragile delicate baby conceived by its parents, carried by its mother, adored by its father ..... and dead. I have a Gambian friend called Manlafi: it means "we don't want it," or "I don't love him" (none of these expressions can be translated exactly into a European language). You might think this is a strange name for a baby? What the name really means, is that his parents had enormous difficulties in having children before, almost certainly with miscarriages or still-births (you see what I mean about translation difficulties). When this son arrived, they were so terrified of losing him like all the others, that they gave him a name to discourage the evil spirits. If the parents didn't seem to care for him, perhaps the evil spirits could be fooled into leaving him alone. Consequently, my friend Manlafi was left lying around, dressed to look neglected, and he survived. The spirits were fooled. There is an important rule of West African social etiquette which every visitor should know: never tell a mother that she has a beautiful baby, for you may bring her baby bad luck by attracting the attention of evil spirits. She already knows the baby is beautiful.

Too often, the spirits claim their own. In some regions, one out of every two children dies before the age of 5; one out of three babies dies in the first year. We read in the Western press about Africa's population explosion, and in some countries the rate of growth is certainly alarming. So many births .... but so many deaths also, and so many unhappy mothers. What is Diallo's brother to do? He must have children. Without children, he is not a real man in this society. Without children, he will not survive old age without misery. African parents are supported in their retirement by their children. No one else will help them.

Diallo's brother has remained faithful to his first wife: instead of divorcing her, he has kept her while taking a second wife to solve the problem. Now he has again been married for five years, and he is still without a child. Poor guy, the ancestors have not been easy on him.

For the wives it is worse. I met my old friend Ibrahim in the street. He is a stonemason, and his black head is now liberally sprinkled with grey. He was cheery as always. I remember so well his marriage problems of ten years ago. At that time, he was wanting to marry the lovely city girl he had met in Bamako, while trying to get loose from the dreadful girl his father had given him in the village. It was Ibrahim who first taught me the saying: "Your first wife is a gift from your father, the second you can choose for yourself." The village girl was a nagger. She could not read or write, and had no desire to learn. Ibrahim was far too dynamic for her. He was too intelligent, too creative and ambitious to live with a village girl. At the same time, Ibrahim was a loyal son and husband: he kept the first wife in the village, and followed as best he was able the Muslim instruction to treat both his wives equally.

I contributed to the costs of the wedding to the second wife he chose for himself, and I enjoyed the food his new bride cooked on the economical wood stove Ibrahim had built in his compound. Ibrahim had become a specialist in building mud-brick wood stoves that focus heat on the cooking pot, and thereby consume one-third less fuel wood than the traditional open hearth composed of three stones.

Yes, said Ibrahim, business was good. He had been building a series of village health clinics, and was very pleased with the new combinations he had been testing with 6% cement, to strengthen cheap, hand-made mud bricks without becoming totally dependent on imported cement. But in his personal life he had terrible problems. "My first wife whom you know, she could never have no babies. It is the tubes which are infected. I have tried everything. I have even sent her to France for tests that they no could do here. But the doctors say they cannot operate. Now she wants me to send her to France again, but I no can afford the money - and anyway if it is no use, then why to waste the money?"

The poor lady probably has venereal disease. Gonorrhea destroys the fallopian tubes. In rural and urban family planning programs, fighting VD is a major priority. In fact, most family planning is now incorporated into what *UNICEF* calls "Child Survival" programs. My wife Jeanne is a militant Child Survival worker: vaccination and hygiene must become universal, infertility must be beaten..... and then we can talk about conceiving fewer children. Family Planning is not just stopping babies: it is part of the strategy to keep children surviving. We will only convince people of the benefits of birth control if we can protect babies from death, keep mothers from miscarriage and infertility. If you knew that half your own children were likely to die, wouldn't you try to have a few more? I know I would.

Maybe Ibrahim carries gonorrhea himself. Many people do in Africa, without even knowing it. After the extreme initial burning discomfort, the body and the penis develop a sort of adaptation to the disease, and the victim assumes it has gone away. But it is still there, and the carrier infects every other partner. In a woman the small amount of urinary irritation from gonorrhea may go unnoticed. The woman can be passing it on, without even knowing she is infected. If she has a baby, she may easily infect her baby's eyes, leading commonly to neonatal blindness unless it is treated at once. Ibrahim's wife has never had a chance to pass it on to a baby. Gonorrhea is very cruel to the fallopian tubes, and Ibrahim's unfortunate wife is probably a victim.

"You see my white hairs, Rob? They are all for my wife. She is threatening suicide. We have tried everything, and I believe I can do no more. Yet I still love my wife. I have just had a baby from my other wife: After eight years without children, you remember that it was necessary to take a second wife, and she gave birth to a boy last month. So, I am happy for that. But my *baramuso* is talking now of suicide." *Baramuso* is the senior wife's title. I promised to talk to my gynecologist friend Fanta, and wished Ibrahim luck. But Fanta merely confirmed my gloomy prognosis. I fear that Ibrahim's wife is doomed forever to the club of African ladies who count themselves failures, unhappy, childless, unfulfilled. Poor woman.

TIME TO FEED THE CHILDREN

One of the best things about being Head of the Family is getting good food. An African father receives the lion's share. In some Bambara families, the Head of Family receives the neck and the feet of the chicken because the women have convinced him that these are the best bits. They do cook them well, but generally in Mali, farmers called Traoré or Diarra or Coulibaly are reckoned to be simple people. If they can be gulled into accepting the neck and the feet, they are simple.

Few West African men are so gullible. In our Fulani family environment in Magnanbougou, the sacrifice of a chicken brings to the Head of the Family the wings (both of them), the heart and the liver. Among the Tall clan, those are reckoned to be the best parts of a chicken. The person who slits the bird's throat (it has to be a Muslim man) receives a specific share, commensurate with his power, and the relationship which he had with the dead bird: before you kill an animal, you normally speak with it and tell it why it is about to be sacrificed. The killer gets the head, the neck and the feet. Those are tasty and nutritious, as well as symbolic. The carcass goes to the person who prepares the stew: probably the mother or the grandmother. That leaves the breast, the legs and the parson's nose.... for the whole of the rest of the family.

You may of course agree with me that the family actually gets the best bits. That's not the point. The point is: there are probably a dozen of them. They don't get much of a share each. Usually, the family sits around a large enamel bowl made in Taiwan or Nigeria. My most frequent dinner companion is a defunct Nigerian President with a chipped nose and a faded military uniform. Inside this bowl is a hefty quantity of millet or rice (in the richer households, where I admit that I eat more frequently, it is usually rice). We all watch as the sauce is ladled onto the cereal base. Then each person mutters "*Bismillah*," and dips his right hand into the bowl. The bigger the hand, the bigger the share. Tiny hands gather but scraps. The smallest children take what they can reach, which is usually shreds of vegetable, with no meat at all. This means they get carbohydrate, a few vitamins, very little protein.

The problem is compounded by the fact that small children must mind their table manners. Funny things, table manners. In Germany, you are supposed to cut your potatoes with a fork: to use a knife is impolite. In Britain you eat bright green peas with the back of a fork. It's impossible, but you have to do it all the same. Most people don't even use forks in West Africa. Spoons are popular, but using your fingers is better.

Spoons or hands, good manners in West Africa say small children are not supposed to reach into the center of the bowl. If a small hand strays too far in, it receives a smack. The meat is usually in the middle. If an indulgent elder distributes pieces of meat in their direction, the small children may get a piece. But if nothing falls in their patch, they are not allowed to go digging.

One of the most important messages that people like Jeanne are promoting in the villages is therefore: "Before you serve the family, set the infant's share aside in a separate bowl." If the mother forgets to give the children their bowl, they may not get anything to eat except the millet stodge at the periphery.

Poverty is widespread in West Africa, for sure. But poverty is not the only explanation for child malnutrition. It is the way poverty is shared. If the small children have to take a larger share of the poverty, they die quicker. And you do not have to be poor to lose a child to malnutrition. Plump adults can have thin children, slowly starving on the grains of rice their tiny hands can grab at the edge of the bowl. Let me tell you a story of my friend the African Princess.

When the Princess was a small girl, she was sent to stay with cousins in Ségou, once the proud capital of the Bambara Empire on the banks of the great Niger River. Explorer Mungo Park reached Ségou on 20th July 1796, six weeks before his 25th birthday. After journeying for a year, and covering 1000 miles on foot or donkey, this Scottish doctor finally "saw with infinite pleasure the great object of my mission -- the long sought for majestic Niger, glittering to the morning sun, as broad as the Thames at Westminster, and flowing slowly to the eastward. I hastened to the brink, and having drank of the water, lifted up my fervent thanks in prayer to the Great Ruler of all things, for having thus far crowned my endeavours with success."

Mungo Park was the first European to reach the mythical river beyond the Sahara Desert. Poor Dr. Park was so far gone already that he didn't care anymore: but drinking the water of the Niger at Ségou is a common cause of child debilitation and diarrhoea. Of course, Ségou was a smaller town in 1796.

My African Princess went to Ségou eighteen years ago, when she was ten years old. Her father's extended family lives there, and she was staying with his elder brother and three aunts. Having three wives is less common now in these days of drought and international crisis, but it is an ancient African tradition. I have commented that there are gastronomic advantages in being Head of the Family. Well, in this Ségou family, each wife prepared food for her own children. But she sent the first dish to the *Fa*, the patriarch. So, the old uncle got six chicken wings, two hearts, two livers. Meanwhile, down at the far end of the compound, the little Princess was scrabbling around to get a share of the vegetable scraps, and bit of leg. At home, with a doting father and a single devoted mother, the Princess and her five siblings had plenty to eat. In Ségou she was feeling hungry.

After three days in this situation, the Princess took a radical decision. When lunchtime came, she walked across the compound to where her uncle was sitting in solitary splendour with his three dishes (one from each spouse). She sat down beside him, rinsed her fingers in the bowl of water, and dipped her lithe ten-year-old fingers into the patriarchal dish. He said nothing. Actually, he seemed quite amused and rather taken with his young niece. Of course, a *Fa* would never admit such a thing, let alone show it in front of a mere child. Thereafter the Princess ate every meal from her uncle's dish, while her cousins scrabbled around in the rice and leftovers at the far side of the compound.

What is unremarkable in this story is the fact that the cousins were sharing their malnutrition in the far corner. What is extraordinary is that a ten-year-old girl should dare to take the initiative to cross the compound and eat with her uncle.... and that her parents should have brought her up to be so independent of thought. This younger brother was already moving from the rural to the urban mentality, limiting the size of his family, and feeding his children.

Nineteen years later, that brave and creative ten-year-old African Princess was appointed a minister in the government of Mali's first democratically elected president.

DIRTY WATER AND HYGIENIC WELLS

We were a middle-aged group of armchair experts talking about child malnutrition. We decided it would be better to give small children their food on a separate dish. "It's obvious!" exclaimed my friend Falie as we chatted over coffee in his comfortable suburban house. "If you only stop to think about it, small kids cannot get enough nourishment out of the family bowl. The big ones eat more, therefore the small ones eat less. And above all, they eat less meat and less fish. But somehow we never saw it in the village, nor even when we lived in the town! I guess I was under-nourishing my small brothers for years and years. Amazing how blind one can be. It becomes obvious only when you travel and then you see your traditional life through new eyes." Falie shook his head in wonderment at human short-sightedness and ran his dark hand through his thick mass of Fulani hair.

"Just like I only realized recently that when we wash our hands before eating, we always pass the bowl of water to the small children last. So, they wash their grubby hands in water that is already full of their parents' germs and dirt! It is no wonder they get diarrhoea!"

Now it was my turn to chuckle: I had never thought of that. But that also seems obvious when you do think about it. In Asia and in North Africa, they pour the water over your hands: much more hygienic than using a common pool of mucky water. In Kenya's schools, the development charity *ActionAid* has introduced an appropriate technology called the "leaky tin." It is hygienic and cheap (every village has an old tin with a hole in it). Fill the tin with clean water and hang it on the school balcony. Let the children wash their hands under the trickle of water coming through the tiny hole. Result: clean hands and no cross infection, and all that just for the price of a piece of imagination. Yet Falie and I have for years been washing our hands in the bowl and passing on our dirty water to small

African children. Generously sharing our germs around the family.

Clean water is easy to obtain if you are careful. Dirty water is even easier to get. West African wells are usually supplied with ropes and buckets. The buckets are mainly made from recycled rubber tires, and they leak 25% of their water before they reach the top. The buckets are emptied into a container and then thrown aside. Often the bucket lands in the dirt where animals and children scrabble around. The next person to draw water then infects the well with the dirty bucket. A good well must have a concrete apron around it and a Village Hygiene Committee responsible for making sure that the area is kept clean. We are always telling rural development workers: "Digging wells is easy, but it is not enough. The well is not the solution to health problems, only a means. If the well is infected, how have you helped the villagers? We have to go further than just digging a well. The solutions are providing clean water, and teaching better understanding of basic hygiene."

German well-digging programs provide efficient concrete aprons topped with a hand-pump (and often a sign in Arabic telling us that the Saudis paid the Germans to dig the well). But the Germans arrive with their equipment, their boots and their beer...they dig...and they go home. Technical specialists limit themselves to the technical business, like mixing concrete. Well-diggers need to join forces with hygiene educators to ensure that their wells actually have a positive impact on community health.

Even when a well is clean, how can we be sure that clean water will reach the consumer? Water is usually drawn by young girls, transferred to a metal basin for carrying, then poured into a clay water jar for storage. People dip the same cup into the water jar that they drink from. This is a universal and dirty habit. A cool water jar is a pleasant place for germs to live and to proliferate. When I am in a village, I drink water from the well or from a pump. I try to avoid water jars. Yet it would be so easy to filter water, to disinfect jars and basins, to avoid contamination by using one cup for drinking and another for dipping into the water jar.....

I was leading a group of extension trainees on an evaluation mission in Kolondiéba, a Bambara village on the Ivory Coast - Mali border. As I came around a corner in the village, I noticed a well with a fence. That was unusual. I walked over to the fence, which was made of neatly trimmed dead-wood branches. Between the fence and the well was a carpet of clean gravel. Most impressed by this sign of good organisation, I went inside the fence to see the well construction. It was built with "Dutch bricks," a strong lattice-work of locally-made bricks which gives an attractive modern look far better than the traditional unlined wells. The well had obviously been sunk by local well-diggers, but with supervision from one of the voluntary organisations, almost certainly *Save The Children*. My reverie was interrupted by a shout.

I turned to find a polite villager standing at the entrance to the well enclosure: "50 francs, please," said he. "With pleasure" said I, fishing in my pocket and wondering why he needed 50 francs. "Thank you. That is the fine for walking on the gravel without taking your shoes off." My jaw dropped! This was an amazing experience. The village hygiene committee had established careful rules about the well, and I was a transgressor. I paid my 15 cent fine with pleasure. Would that all villages were so cleanliness-conscious. On further discussion, it turned out that this was a triumph of clean extension work by *Save the Children* and the villagers were taking this hygiene business very seriously indeed.

At that moment my trainees ambled into view. "Come and see the well" I called.

"But first you all take off your shoes." 25 people obligingly pulled off sandals and flip-flops and lined them up neatly along the fence while they came to admire the Dutch bricks and congratulate the villagers.

I dug into my pocket again, and pulled out a bill of 500 Fcfa ($2). "This is the fine: 50F for my mistake and extra money for friendship and solidarity," I grinned. The hygiene policeman immediately called over to the occupants of the nearest house and within seconds there was a knot of villagers waiting to witness me pay my "fine." The trainees gathered around and collected their footwear while I gave over my contribution to the hygiene fund. A praise singer started to screech about the benefits of the well. The hygiene policeman waved the 500 francs around over his head. Two village women started to dance. The trainees joined in with rhythmic clapping. We had turned a fine into a party. By now there were nearly a hundred people gathered around the well, and everyone was enjoying a spontaneous celebration of water hygiene. That's how effective extension work is done in rural Africa.

FIRE AND AIDS, SEX AND VENEREAL DISEASE

The telephone rang. It was my dear friend Fanta the gynecologist wanting to report on the case of a driver's sister. "She has a prolapse of the uterus, poor lady, and she is very uncomfortable. But I'm afraid I cannot operate on her until she has got rid of her syphilis. So if you could get that treated first, Robin..... She has a prescription for procaine penicillin, and she must have two tests to show that she has cleared up the disease. If I operate before, then I might catch syphilis myself: and no thank you!"

Venereal disease is a general problem in West Africa. Up in the Sahara Desert they say the nomads are born with it. Syphilis is probably the major health problem among the Tuareg, the famous "blue men of the desert." Down in the cities, gonorrhoea is more common, but syphilis is endemic. Studies in Yaoundé (Cameroon) referring to the late 1970s show that for professional prostitutes the rate of infection for both diseases was 16%. This suggests that between 16% and 32% of prostitutes are carrying one or other of the diseases, or both. It must be worse now in 1991: after all, 60% of prostitutes in Bamako are thought to be HIV positive, and one would expect higher AIDS figures down in Yaoundé than in Sahelian Bamako. 22% of pregnant women are infected according to samples in Blantyre and Lusaka: although we must note that these are urban samples, in high-infection countries, and therefore not typical of Africa as a whole.

"There seems to be a link between the gonococcus and the AIDS virus," says Fanta. "I cannot explain it: but the same people seem to have both, especially among prostitutes. I think that more than half the prostitutes have gonorrhoea. But it is very widespread among the general population. There is a lot of self-treatment, so you can't measure infection rates: but in my clinic about 60% of patients have gonorrhoea or syphilis."

Syphilis is more insidious: more difficult to spot, more difficult to cure. It gnaws away at you, and you may not even know that it is there. Gonorrhoea is easier to catch, easier to see, quicker to cure.

For men especially "a dose of clap" is obvious. It hurts. You cannot miss it. West Africans call it "*chaude pisse*" in French, and in English they call it "*fire*." Among young prostitutes, or "apprentices" to the profession, the infection rate in Yaoundé was 52%. So as a rule of thumb, the younger and less professional your partner, the higher the chance of catching FIRE!

This is of domestic relevance to me since I have just been obliged to inspect my gardener's dingle dangle. Leo took me into a corner, and told me he was ill. "Dis time eet ees not ze malaria" he said mournfully. Then he pulled down his trousers and showed me a drip of pus. "Ça fait mal." I believe him: the FIRE of gonorrhoea is renowned for the pain, at least during the first week. Leo is walking around with his knees apart, looking distinctly miserable.

Normally in these cases you treat both partners. But Leo cannot even remember where he met his partner, just that she was young and that he gave her the price of her taxi fare to get home. Perhaps the statistics have changed over time, and through geography: but if it were Yaoundé I would put this girl plumb into the "52% of apprentices" bracket. I'll go off later to buy the procaine penicillin.

As Fanta indicated to me, there is a school of medical thought that now links the rapid spread of AIDS to the huge rate of VD infection in West Africa (with the notable exception of Nigeria, which seems to stand out on the AIDS map like a beacon of cleanliness). If VD lesions are very common, and if the AIDS virus can spread more quickly through those lesions (particularly syphilis and cancroids), would that not explain the fact that AIDS has been spreading like WILDFIRE?

FIRE may be immediately more uncomfortable when you get it: but AIDS is an infinitely greater threat to personal health. In some countries it is getting way beyond a personal health problem: it is almost a threat to national survival. One in 40 African adults is thought to be infected (world figures 8 million HIV infections: 1 in 300, which is still a hell of a high ratio). The World Health Organisation estimates that 500,000 out of the 800,000 adult AIDS cases worldwide, have occurred in Africa. So, with less than ten per cent of the world's population, Africa has well over half of the adult AIDS sickness.

Given such figures, Leo's sore penis seems a pretty feeble thing. Even syphilis pales in significance (though I'd still rather not catch it). One characteristic sight in the African slums, is seeing elderly people rocking from side to side as they walk. They walk with difficulty, on a wide base, which is a typical syndrome of advanced syphilis. It the future these patients may disappear from the medical landscape: not because of medical progress, but because AIDS will take them off before their syphilis infection can become advanced. It is a sobering thought.

Far worse are the 1991 children's figures: nearly half a million cases, 90% of which are in Africa. The WHO now believes that as many as 10 million children may be born HIV positive by the turn of the century. While Jeanne and her charitable non-governmental colleagues are fighting child disease in the villages alongside the under-equipped and underpaid government health services, there is another little agent working away to increase child mortality: the AIDS virus. Jeanne and Co. are reducing neonatal tetanus, improving hygiene, reducing death from diarrhoea....and death rates are falling. For how long? WHO reckons child mortality will soon be on the up again, by as much as 50% in some countries. The Sahel is not threatened anywhere near as much as central and southern Africa. But the problem will still be with us.
There will be lots of children with sick parents or with no parents. If we discount those children who will die from AIDS, there will still be 10 million orphans in Africa by the late 1990s,

according to UNICEF. Who is going to grow the food to feed them? We are heading for a new Africa, where the population explosion will be a memory of the past..... where elderly grandmothers struggle to feed and clothe a dozen or more grandchildren. In Uganda one estimate puts the number of orphans already at 1 million (out of total population of 20 million).

In the next century perhaps we shall think back to those far off days, when we were all worried about the population explosion, and whether we would be able to feed the people. Will we then look at the orphanages surrounded by empty fields, and still wonder how to feed them? Will we then long for the days when the towns and the villages were full of vigorous adults with *chaude pisse* but nothing worse? And then look around at the misery of hungry children marauding in bands, with no parents to guide them?

The Black Death in 14th Century Europe cut into the workforce, and had the result of raising dramatically the wages and living standards of farm labourers. In the 1990s the most lucrative jobs are probably customs officer, smuggler, army officer, marabout holy man. In fifteen years' time the best wages may well be paid to doctors, undertakers, and farm labourers.

YOU WILL NOT EXCISE MY DAUGHTER

Female excision is a hot topic among Guardian readers. I am against female mutilation of any kind. The World Health Organisation is in the anti-mutilation business too. They have made a film about excision in Burkina Faso, with the above title. Well done the WHO! They ought to get it shown and repeated, and re-repeated, on every television station in Africa and the Middle East. Here's what happens in the film.

A young urban family from Ouagadougou (the sprawling and dusty capital of the Mossi homelands which used to be known as Upper Volta) makes the monthly visit to the grandparents (in this case the husband's parents) and the urbanites find themselves up against tradition. The family arrives and says "hallo." The calabash of water (I hope it was millet beer) passes around to show they are welcome and the usual litany of greetings is exchanged. Now the wife has gone off to see Grandma in the kitchen/the compound and the kids are racing through the dust with the cousins, chasing birds away from the millet crops. Grandad is sitting on a low stool in the shade of his thatched house, talking man-to-man with his eldest son: "I have to speak with you about a serious matter, my son. Your daughter Nafi has passed the age of excision. The old women and the fetishers are annoying me. I am the chief of this village and yet the elders are criticizing my family. We must therefore organize Nafi's excision without delay."

The Chief's son disagrees. As it happens, he is not only a modern man (one wife, two children, city clothes, a small clean house with a video, and a Peugeot car in which to visit the village), he is also a medical doctor. The doctor traces all the health arguments and tells his father that excision is out of the question. When the Chief tries to come on heavy and insists on excision, his son just walks off, leaving the old man fuming.

The village is divided, and everything that happens increases the divisions. The doctor carries out his free monthly clinic as if nothing had happened. Meanwhile back in the kitchen, the doctor's wife is grinding flour with a stone roller, such as were used by our stone- and iron-age ancestors. In the African village, labour-saving technology for the housewife is very rare. The Village Council, under the chairmanship of the Chief and the animation of his son the Doctor, discuss whether they should now invest in a village pharmacy and how they can reduce the population's major health epidemic, malaria.

Having settled the matter of better health, the Chief returns to the important matter of tradition...... and excision. "As long as I am Chief, my decisions are commands" thunders the Doctor's father. The Chief's official tam-tam drum echoes the Chief's official words. Several elders speak in support of tradition. Then an old soldier steps forward resplendent in grey beard and medals.

"Who are you to attack Doctor?" he cries. "Bakary, I didn't hear you complaining when Doctor operated your hernia last year. As big as a football it was! Which of the elders here doesn't remember Bakary's hernia? Why Bakary, you could hardly walk last year. You had to hold your hernia in your arms to stop yourself falling over. But you weren't criticizing Doctor's ideas then, were you? And you, Yacouba: what about your goitre? The size of a Baobab fruit it was! Before Doctor removed your goitre for you, you had a voice like a toad, and your only contribution to the village was croaking pleas to Doctor to help you. Now that he has given you back your man's voice, you use it to attack his ideas. Is that gratitude? Is that wisdom? Which of us had Doctor's knowledge about the goitre? None of us!
 And which of us has Doctor's knowledge about excision? If he says it is a dangerous practice, who here has the knowledge to refute him? So, we should recognize that new knowledge is a good thing, and we should listen to Doctor's information about excision."

The Doctor gave a presentation in the middle of the village under the Palaver Tree. With flip charts and photographic enlargements, he presented the horrors of excision. The villagers watched in stunned silence as Doctor showed pictures of forbidden secrets and hideous deformities. He spoke of girls dying of tetanus after being excised with a rusty blade, of girls bleeding to death, of urinary fistulas making women incontinent and unsocial for life, of childbirth pains and deaths, of frigidity and fear and divorce. "Circumcision for boys is normal, necessary and hygienic," finished Doctor: "But no religion, no prophet, no doctor has ever recommended excision."

The following month he brought in a panel of ethical experts: one Islamic imam, one Protestant pastor, and one Roman Catholic priest (known in Burkina as a *mon père*). The panel members were unanimous that excision is a thing of tradition, not of religion. The pastor said it was forbidden; the *mon père* condemned it as diminishing the dignity and status of women; the imam stated that it is "not required by Islam." After a moment of silence, the Chief declared that he had heard their views, but that "Tradition is not the same as Religion." Confronted by such ethical solidarity however, the Chief now ordered his fetishers to go off and consult the Ancestors, the spirits of the Rivers, the spirits of the Bush, and the spirits of the Mountain.

While the Ancestors pondered, Doctor was taking his campaign to the women. In a question-and-answer session chaired by a midwife, he heard all the fears of women who had, as he said, "suffered from this servility invented by man to guarantee the fidelity of women." Is it true that a woman who is not excised will die in childbirth? "No, that is not true: the midwife here can attest to cases of unexcised women who have given birth without problem." Is it true that a baby will die if it touches the clitoris? "No, that is superstition quite unfounded in truth." Is it true that a child born to an unexcised woman cannot be baptized by the imam? The doctor explained what had been said by the ethical panel and the imam in the Village Council:

Islam does not recommend excision. Yes, a child can and should be baptized; no, it is untrue that such a child will be denied entry into Paradise; and if your mother-in-law says that she will refuse to carry hot water for you when you give birth, this is because she is still under masculine domination and you must reason with her.

Back in the Ancestor House, the chief fetisher was offering a white cock. "If it dies on its back, let the answer be "Yes".... that excision of women is strong custom which should be maintained. And if it dies on its belly, let this be the answer "No," that excision is unnecessary." The fetisher slit the throat of the hapless cock and spread the gushing jet of blood across ancestral altars strewn with feathers from countless previous sacrifices. He dropped the cock. It raced around the floor like a headless chicken (which it had now become) until it dropped, shuddered, and lay still: on its belly. The fetisher's jaw dropped open. "But if we stop this tradition, he groaned, our influence will be gone. This cannot be. Tradition must be protected. We must deny the ancestors, and maintain excision." And thus it was reported to the Village Council.

Doctor and his wife went back to Ouagadougou, leaving the kids with Grandma for the week. And that very night the word went out that the old women should excise three girls—including Nafi. Fortunately, there is very little secrecy in a village and some youths overheard the news. While the old women proceeded to the Excision House, two young men went racing off into the bush on a "mobylette," one of those ubiquitous little Peugeot mopeds which make the citizens of French-speaking countries so mobile. Nafi was to be the third of the victims. Even as she was dragged screaming from her grandmother's hut towards the bloodied ritual knife, a police van raced up and all the old women were arrested. For excision is against the law in Burkina. As Nafi was led home shaking, a policeman bent over another of the girls, lying curled up under a blanket. She had bled to death. Her body was still warm, warm as the pool of blood soaking into the earth at her side.

The following day the film finishes in sorrow and in victory. The village is in mourning. The dead girl was the only daughter of the Chief Fetisher, he who defied the Ancestors. Today it is the men who proceed to the Excision House. There they set fire to this hut of shame, and the Chief declares that his doctor son was in the right. "As long as I am your Chief, my decisions are commands." Once again, the ritual tam-tam drum echoes the words of the Chief. "I have decided that excision is unnecessary, cruel, and not desired by the Ancestors. If it ever happens again in any family of this village, I shall personally castrate the head of the household."

THE EYE OF YOUR NEIGHBOUR

"IN THE OLD DAYS," mused an old friend from the Niger interior delta, "it was the eye of your neighbour which kept morality in good shape. You could not steal and get away with it, for the neighbour would see and you would be humiliated." My friend is a distinguished Malian national figure, revered for his age and his honesty. "I live in a small house that I was able to build during my career as an administrator. When I see the concrete palaces built by modern politicians, army officers, civil servants...... Well, I ask you, how could they build such mansions with their salaries? ..... When I see their huge buildings going up, then I am proud of my small house. It is a small fruit for years of effort: but it is honest fruit, and I grew it myself."

He removed his round white hat and scratched his grizzled crew cut. "You know, it may be because I am a Fulani that I think like this. In our culture we say "He is the son of X"...... and this is enough to make a son keep to a strict morality, in order not to shame the name of his father. But more even than that, I believe that morals are lax because of the urban environment. In the village, people are kept in a good morality by the eye of their neighbour. In the town, those rules are lost."

Fulani nomads accumulate cattle, so their idea of "wealth" has an especial connotation. They used to accumulate slaves too, to cultivate their conquered lands. Each Sahelian society works differently. In the villages of the Manding peoples, there are rules that limit the possibility for anyone to become too wealthy. A traditional Village Chief cannot sell his land, nor can the villagers: the land does not belong to them. Land belongs to the whole community together (including to those who have not yet been born). Land for farming is allocated by Elders who are descended from the founder of the village – he who first cleared the land and established a village. Traditionally in

Africa, as in America before the arrival of White European settlers, land was not privately owned.

The Chief may protect the interests of the founding clan, but he cannot easily make himself personally rich. Nor is it really possible for the Chief to enrich his sons or to found a dynasty: for although the chiefdomship is generally hereditary, In Mali it passes not to the Chief's son, but to his brother. Or at least it passes to one of the senior members of the lineage in the oldest generation: very rarely to the son of the deceased Chief (and even more rarely, let it be said, his daughter). So if a Chief has enriched his sons at the expense of their cousins, for example by allocating to them the best pieces of arable land, it is unlikely that they will keep their advantages.

As in many societies, capital accumulation is limited by the Manding village social system. If you have more resources than anyone else, you will receive more dependents to feed than anyone else. When one household is suffering hardship, loses its breadwinner, or cannot harvest due to illness, everyone else in the community offers support; this is not "communism" but it is a form of "communitarism" that tends to protect families according to their needs, with assistance from each according to their means. If your brother dies, you must feed his children (and probably you will marry his wife as well). In this way, a village remains socially cohesive and economically balanced.

The same social system extended to the town becomes a disaster! Whereas in the village, extra mouths also mean extra hands to collect wood, to hoe crops, to pull water from the well, in the city they mean extra people to feed on a salary or wage that remains static. Far from having extra hands to collect wood, more dependents in the town mean that you must buy more wood from the fuel merchants to cook more food—and all this from a single income, which may be insufficient anyway. In the urban peripheries of Africa, you find misery. Whatever economic activities people can find in the urban outskirts, they are usually sharing and circulating poverty. There is very little wealth creation among new urban immigrants.

The extended family system provides a wonderful support mechanism for weddings and funerals, for looking after children, and caring for old people. But when people are not well off, it can also be a mechanism for spreading poverty. Young couples starting off in life find themselves swamped with nephews and nieces, feeding, clothing, and schooling dependents who consume the money they were saving to educate their offspring, buy a car, or build a house.

These are pressures that force civil servants to supplement their incomes in dubious ways that I call "daylighting" (as opposed to Western European "moonlighting"): like teachers selling good exam results or passing only those pupils who pay them for private tuition. Some civil servants go further than "daylighting" for survival. Is it very wrong to over-invoice the government, If you need that to feed your children? Survival-corruption can be understood; but too many African officials go on to amass illicit riches. Why so? It is because the anonymity of city life allows you to behave badly "out of sight of the eye of your neighbour." Indeed your neighbour in the city is probably also accumulating illicit wealth as hard as he can.

The nature of wealth changes too. In the village, everything belongs to someone you know. If it is his, you cannot take it. But in the town, everything belongs to strangers. Your urban neighbours are not kin from the same village: they are strangers. If you have an opportunity, why not grab it? In the village, wealth, food, and cash are all earned with the sweat of your brow, or with the sweated efforts of your kinsman. It is "warm money." But in the town, wealth is anonymous, it has no sweat; it is "cold money." Cold money is anybody's money. And no money is cooler than the anonymous money that comes from banks, from the government, or from the white colonial traders. If you don't take advantage, the stranger next door will because he is seeking to survive. Or he is seeking to get rich? How do people get rich "out of sight of the eye of your neighbour"?

"Pele" is a trader. Naturally enough, with a nickname like "Pele," he used to be a good footballer: good enough to play in the national side. Now he has a company called "General Trading," and that is what he does. He looks around for opportunities, government contracts and suchlike, offering good prices and cutting deals. He does a lot of office furniture. A "big brother" of Pele (not same father, not same mother) working in the public works ministry just helped him to get a good contract to supply desks and other stuff to the ministry. Pele asked his "big brother" what he could do for him in exchange. "Nothing at all, little brother," replied the Big Man. "What I like to eat, you cannot provide. So, you needn't bother to give me crumbs."

I confess that I didn't have a clue what Pele meant by this story, so I asked him. "The Big Brother gets big contracts," explained Pele. "In the Ministry of Public Works, contracts for roads and bridges are huge. When Big Brother negotiates ten per cent from a building contract awarded to a construction firm, he can win 100 000 dollars right off. That is what he likes to "eat." Of course he does not eat all of it: the minister and other senior officials receive a share. In the cost negotiations with foreign firms, they build in the ten per cent right there. But if I give up ten per cent, there's nothing left from the sort of small contracts I go for. He knows that, so he is leaving me with my profits."

As usual, I encouraged my friend to reveal a bit more. He explained how he works when there is no Big Brother. "I have a system. I go straight to the senior ministry official and tell him that I wish to begin a social relationship with him. If I have it, I will help him; if I do not, then I will tell him that I cannot help him. And with that, I put a note for 10 000 Fcfa ($40) on the table and ask him to take it as a sign of friendship. If he takes it, he will help me win a contract if he can. He knows I will give him something if I win the contract thanks to him. Another time, if he needs something, he can telephone, and if I can help him, I will."

It was Friday, and Pele was dressed in an embroidered gown ready for the mosque. Usually as a businessman, he likes to wear a suit and tie and he is always pulling at the tight collar around his muscular neck. He was famous for heading the ball in front of the goal posts, which gave him neck muscles that do not easily fit a tight shirt collar. He looked relaxed, sitting back waiting for the call to prayer from the muezzin.

"I do not like to wait until the man asks for a present, for he may ask too much. So, I use my system and it works very well. But some people do not know how to offer presents. Do you know? I was once present at the official opening of bids for a government contract. This took place in front of the bidders and the financiers and about fifty people were there in the room. The Head of Department opened this envelope and inside there were two sealed bids: the first envelope carried the words "BID FOR CONTRACT." And the second envelope was inscribed: "$200 for the Head of Department from X." Everybody in the room started to laugh. The Head of Department handed back both envelopes to the bidder and he left in confusion."

It takes a lot to embarrass these greedy traders or indeed their venal cronies in government service. But on that occasion, there were simply too many eyes of too many neighbours.

WARM MONEY, COLD MONEY

The World Bank report complained that Mali's Government Rural Development Organisation was getting only 20% repayment from peasant farmers for their loans. The report's author pointed out that the nearby Oxfam-funded program made smaller loans and that their repayment rates were 98%. He wondered why. I believe the answer lies in the following explanation from a Mossi farmer in Burkina Faso: "Warm money is what you earn yourself, when you have gained it through the heat of your work and the sweat of your body. Cold money comes from outside. Cold money does not belong to us. Warm money is important, it is for you. Cold money belongs to other people."

"Thus" says Canadian Professor Guy Bedard who publicised this Mossi explanation, "the peasant explains what he thinks about loans. If the money is warm money, the peasant farmers will look after it, will make it work for them, will make sure that it circulates, and will make sure that it is reimbursed by applying appropriate social pressures. But no one bothers to look after cold money."

The Oxfam-funded program encourages cereal banks along the bend of the Niger River in North Mali. The farmers sell grain to their own cereal bank, and buy it back again when they need it: either for seed in June when they plant, or to eat in July and August when their food stocks are very low. They can borrow too. If they borrow one bag of seed in June, they repay 1.5 bags after harvest in October. 50% interest in grain is so very much cheaper than the 150% in cash they have to pay to a money lender. The farmers decide on the rate themselves in a General Assembly, where every man can speak. I was there when the assembly outvoted their committee on the interest rates. The committee proposed repayments of 1.25 bags.

"No" argued the members: "With storage and weight losses, our cereal bank will not make enough profit. We must pay more!"

So, the decision was taken by the men themselves. Not the women. Songhoy society is not that evolved - or rather I should say that Songhoy society just doesn't work that way. But the idea that each family has a voice and a vote is already a major step forward. They are deciding about their money, and this makes the whole operation very warm.

Anybody working in village development has to deal with credit. Subsistence farmers don't save money. Many of them don't even use money. If they want money, they sell some groundnuts or some rice to buy medicines, or maybe a Taiwanese cooking pot, a Nigerian enamel basin, or a Japanese transistor radio. Their only other use for cash is to pay taxes to a government they don't need; or to pay fines to officials for breaking rules they've never heard of. Any farmer who can get his hands on "cold money" through a government or World Bank credit scheme is simply getting back a share of his taxes. How wise of him to hold onto it, rather than letting some civil servant steal it back from him by calling it a loan.

What is a loan? You will find that a difficult concept in Malian villages where everything is owned by the family. If I have a radio, my elder brother can come and take it at any time. The rule is this: if you have it, you must share it. If I have no money when someone asks, then they will shrug. If I have some, then they expect a share if they are in need. This is not a loan: it is a share. So, if you are asked for a book or a hat, do not expect to get it back. If you ask for it back, you will probably find that it has moved on to a cousin. Ownership of 'stuff' is a principle much loved by White people, not so much by West Africans.

Yet in changing Africa with its modern cash economies, farmers do buy things and they do use cash. They need to buy modern conveniences, like better steel plowshares, improved drought-resistant seeds, medical facilities, and education for their children. As farmers serving an evolving market, they need access to credit like any other farmer or businessman. If they can get hold of some cold money, so much the better; but there is not much cold money available in the remote areas. So small-scale village schemes work with warm money. They make warm money work. When they get access to warm money, the farmers usually look after it pretty well. Traditionally there isn't really any savings system. If you are a man in need of cash, and if you have no more spare crops to sell, then you have to seek help from "relations" or "money lenders."

The traditional form of saving for women is so warm it is almost cuddly—a mutual savings club called a "tontine." You can find them across West Africa, in town and countryside alike. Awa who lives with us (and whose small daughter Baby Jeanne is named after my wife Jeanne) belongs to a savings club of seven women. Awa and her friends each pay 100 francs (30 cents or 20 pence) per day into the tontine. They never forget and they seldom fail. If they really and truly cannot find the money today, then they can make it up during the week... or more likely, one of the others will chip in an extra 100 F to keep the friend's and the group's payments up to date. There is immense solidarity within the group of seven mutualists.

Every month, one woman takes the pot of 21,000 Fcfa ($75). The first round is drawn by lots, and thereafter they each get the pot in turn. You can ask another woman to let you take her round, if you are in need. Thus, for example, Maladon's child was sick last month and she was faced with a big medical bill. Awa agreed to give up her turn, so that Maladon could get the 21,000 this month and cure her child. Bad luck on Awa who had calculated on her money to buy clothes for the children and herself for the Eid festival, but Maladon's need was greater than Awa's. In Africa solidarity takes precedence over economics; medicines for a child take precedence over dresses for a fete.

That is the secret of the tontine: it is not really an economic savings club so much as an insurance fund, a "mutualist" system. Such clubs existed during the Industrial Revolution in Europe to guarantee workers a decent funeral. The miserable poverty of life should not be sealed with the indignity of a pauper's death, so workers paid into a fund that would guarantee them a respectful burial. In Rwanda a similar "death group" mutualist system exists to provide a good Christian wooden coffin in that most Catholic of countries.

In some countries the tontines have expanded in the modern economy and have even begun to take on an economic role in the formal economy. Cameroon's tontines are more powerful than the banks. The tontine is warmer and more reliable than a frigid foreign bank, more predictable and more flexible than the rigid banking system with its imposing buildings and men-in-suits. Nothing makes money colder than a self-important official in suit and tie! "What's the definition of a banker?" asks my smarty-pants 14-year-old. "Answer: Someone who only lends you money after you have proved to him that you don't really need it."

The mutual savings banks (like the Banques Populaires of Rwanda that were so successful before the genocidal 'Evènements' that killed 1 million people) combine the warm intimacy of the tontine with the economic potential of the banking system. While most banks follow the tradition of lending to the rich, Banques Populaires collect the savings of the smallest peasants and lend it back to them. 83% of Rwandan borrowers were peasant farmers. In this small country which the World Bank counted among the poorest in the world in the 1990s, the Banques Populaires mobilized $25 millions in their first 12 years of operation. There were 95 Banques Populaires in Rwanda and 52% of Rwanda's 6.5 million people were members until the fighting broke out in a Hutu racist campaign against Tutsis.

Mutualist savings banks have been a growing sector in Africa for 25 years. People believe in these village-level savings banks because they can see them, feel them, smell them, and know them. The committees are composed of local people and they decide who gets what loans. If you need a loan, you can talk to someone you know—someone who understands why a marriage is more important than a wheelbarrow. They know when you have to help your sister because her child is ill. They're not going to turn you away with a cold stare. They'll welcome you with a warm smile and give you a loan from your own community's warm money.

MARITAL STORMS AND FORGIVING NAFO

Awa has been part of our household for ten years – in Mali, then in The Gambia, and now back in Bamako. She does some washing for us and we feed and clothe her children. The eldest is Nana, aged twelve; then comes Ousman Robert; and, finally, Baby Jeanne, named for the white friends in whose house they were born. Their father Nafo comes and goes. A few weeks ago, he came back from the Coast, with some new city clothes and a nice transistor radio: meagre profit from six months working in Banjul. We were not even certain that it was to Banjul that he had disappeared. One day, we heard from Awa that he had gone somewhere. So, we rescued Awa and brought her home (with her kids, of course).

Nafo is rather moody. Having abandoned Awa and the kids, he is both relieved and disappointed to find them getting on well without him. He has been disagreeable these past days and very sweaty, because I suppose he feels he has nowhere to wash. He could wash in our garden, but that is where Awa is living. Awa has not made him welcome since she feels he went off and abandoned her in Bamako. Indeed, that is exactly what he did.

Afraid he might start beating her, Awa went off to hide with an old woman neighbour. Nafo came to complain to me. Apparently, Awa had dropped Ousman on the way and sent him to Nafo's family because in West Africa, "children belong to the husband's family."

"But she did not do according to our custom. She left him at the bottom of the road and told him to walk to my brother's house and tell them that he was arrived," explained Nafo. "If she wanted to divorce properly, she should take the children to the house of my elder brother, put their hands in his hands, and tell him, "here are your children," and then she can leave and we are divorced. She may not know our custom correctly because she is Bambara and we are Senoufo, but that is the correct custom for us."

Since they are not legally married, I sort of wondered how they could become divorced. But I let that detail go.

Long debates ensued. When Awa came to do our family washing, I insisted that she stay, and listen, and talk. Three times Nafo fell on his knees in front of me, hands behind the back, to beg forgiveness by touching his forehead to the ground. I said: "That is all fine, but you've done it too often before, why should I or Awa believe you today?"

"But if Nafo begs forgiveness in this way, you have to excuse him. And then you have to instruct Awa to return to Nafo's house. That is our custom," said Baboucar Diarra, a wise fellow of mature years whom I had asked to come and offer advice.

"And what if he begs forgiveness again and again, but he always falls back into his evil ways?"

"If he begs forgiveness of the wife's relations, then they must forgive him. If he does it twenty times, they will forgive him. If they are convinced that he is not serious, then they will impede him physically to stop him falling on his knees to beg forgiveness, but if he is there on his knees, then they must forgive him. That is our custom."

Boubacar is erudite, a French-educated friend whose advice is precious. He is Bambara like Awa, he is wise, and I am bound to accept his advice because it is I who asked him to help. His advice is approved by the rest of the group: assorted drivers, cooks, gardeners, and neighbours who have come to listen to the affair and to hear judgement. The family must keep together. The judgement (my judgement as the senior member of Nafo's and Awa's "family ") has to be compromise. Awa and I must allow Nafo to prove himself again.

There is quite a lot to forgive. In one of his spates, Nafo gave me a long rigmarole about Awa and her promiscuity. According to him, Awa has had 6 husbands, two of them leaving her with huge debts from her mismanagement which he, Nafo, gladly agreed to pay off. None of Nafo's accusations ring true. Nafo himself has so many debts, how could he be paying off a clutch of previous husbands?

Nana is not his child in this story: "You have only to ask anyone in Bamako who will confirm the bad character of Awa." Not surprisingly, Awa is rather annoyed about this slander. When we first met them, Nafo was the young gardener in the house we were renting and the even younger Awa was carrying Nana on her back. Little Nana was learning to walk – just like our daughter Leïla.

I asked Awa casually who was Nana's father and she laid into Nafo at once, saying that he likes to tell bad stories about her but he is her first and only husband. This seems possible. Just as it is possible that she was pregnant before she fell for Nafo. What does it matter? It is not a matter that has ever worried me in the past ten years: Nana is twelve now and Awa about thirty. I did check with other people, of course: Awa was a single mother selling vegetables in the market with a baby on her back, when she met Nafo and moved into the house where he was living with an older brother and four co-wives. Nana's natural father works as a truck driver.

Awa was glad to move pout of the compound where she was wife of the younger brother, and therefore only number five in the pecking order of wifely seniority. Awa lives with us, in a small house at the bottom of our garden. She has gained both security and independence,. We give Awa no salary, only presents for the children and money for food. Awa has quickly recognized the strength of her position and has brought up this fact more than once: if she has a salary, her husband can demand it. But she still brings home more than Nafo.

As a gardener, Nafo earns 20,000 Fcfa per month (forty pounds sterling). He spends 8,500 on rice and claims to send 5,000 per month to his father in the village (which I doubt). If you add the cost of condiments and cigarettes, and there's not much left over for meat or milk. In Mali, a father is supposed to supply the food and also to clothe the children. Urban ladies spend their money on clothes. Awa likes to be well-dressed. I am not sure whether Awa is spending money on rice, now that Nafo has returned from exile, or whether she is insisting on her right to spend all her own cash on her own clothes like so many wives in Bamako. But we ensure that her three children are well-fed, and we are now sending the first two to school. The family is together for the moment.

SENDING NAFO'S KIDS TO SCHOOL

The headmaster of the local primary school said that there was no space for Nafo's two children. But Nana and Ousman Robert will have no chance in urban Africa unless they can read and write the colonial language. We decided to try the Sogoniko school beside the tarmac road. So off we went one hot Wednesday morning to a double row of low cement buildings with corrugated iron roofs.

When I count the 80-odd children in class 1, I am not optimistic. Mr. Diarra is calling out names. "Marie Diallo." Two little girls in blue arrive and install themselves by the blackboard alongside five small boys fidgeting in their new khaki uniforms. Mr. Diarra sorts out the boys and then identifies the two girls. "You are Marie Diallo? Good, stand here next to Mamadou. And what is your name? he asks the second little girl. She is so shy that he has to bend his ear to her mouth. Aha, you are Marie Diallo, too. Very good. Well, you stand here next to this other little girl." He catches my eye, raises his eyes to heaven in a sign of hopelessness, and grins. For sure, you need patience and stamina with 80 small children in the classroom .... especially when half are called Diallo!

Nafo introduces me to Mr. Diarra. There is no difficulty in getting Nana and Ousman into the first class, provided we are able to pay the inscription and help the headmaster to see his way around the number difficulty. 16,000 Fcfa (32 pounds sterling) is the sum mentioned for the two children: 3,500 each for the official registration, the rest to serve as "motivation" for the staff. I like Mr. Diarra: I prefer negotiating with a clever rogue, far better than with an honest but stupid man.

Mr. Diarra is soon back assuring us that he has fixed it. All that remains is to purchase the uniforms (three for Nana, two for Ousman), their desks, the school books, the notebooks, the pencils, and the writing-slates. Ah, one more thing: "The children can sit on the floor today and start right away, but they must not come to school, please, on Monday. On Monday the Minister of Education is coming to the area for his annual inspection. The school must remove surplus students before his arrival. The children can return on Tuesday and bring their desks please, on Tuesday. Thank you so much."

The minister is a military man. In one area he found 130 kids in a single first-year classroom and there was hell to pay! In Sogoniko, there will be no more than 40 or 50 in the class on Monday. "Business as usual on Tuesday, back to 80 or 90 in class." Nice work, Diarra!

And thus it happened that Nana and Ousman Robert went to the Sogoniko school. For three days they sat on the floor. Then they had a couple of days off until the minister had passed by. And, finally, they arrived at school one Tuesday morning in uniform (Nana in blue, Ousman in beige), clutching their slates, and carrying the wooden desks I had ordered from a local carpenter.

Two weeks after the beginning of term, I heard three claps outside the door: African doors are kept open and do not have knockers, so you clap your hands to announce your arrival. I stepped outside into the hot sun, to discover Nafo and Mr. Diarra, come to explain the problem of Nana's birth certificate. Nana is too old for the first-year class. She cannot stay in school in first grade if she was born in 1979 or 1980. We cannot change the rules. But there is a solution (In Africa, there is always a solution). The solution will be to change Nana's age! There is in the ministry an amiable gentleman who produces "duplicate" birth certificates to order. In this way Nana can appear to be a couple of years younger, and she can stay at school.

The birth certificate will cost around 1,000 Fcfa; then we must add "motivation" for the amiable gentleman, and "transport" costs for Mr. Diarra, who alone can make such convenient arrangements; and then there is an additional little something for Mr. Diarra's trouble: all in all, another 6,000 Fcfa to add to the cost of living in West Africa.

Here is the school account:

School registration and bribes	16,000
3 uniforms for Nana	6,000
2 uniforms for Ousman	4,000
2 desks	3,000
1 copy of the school book "Binta et Mamadou"	3,500
2 exercise books	300
2 pencils	200
2 slates	400
false birth certificate and bribes	6,000
TOTAL	39,400 Fcfa =

78 pounds sterling (around $90 US)

A watchman/gardener like Nafo may earn less than 20,000 Fcfa per month. The salary of a senior official in a Malian ministry may not reach 80,000 Fcfa. If he has only two children, his whole salary will go toward school. Most people here have more than two children, and they are struggling. Did somebody in the political establishment say that schooling is free here? In that case they are quite ignorant of what life is really like down there on the city streets of West Africa.

Postscript in 2024: Nana – now a strikingly beautiful in her forties, with three children - enjoys having two birth certificates. She often asks me in Bamako, wearing a cheeky smile on her attractive face, how old I think she really is.

EDUCATION FOR EXILE

"Salif has called to see you." Salif? Never heard of him! Yet another supplicant, I suppose. I get this sort of unexpected visit three times a day. Twice a day they are people I know, the third is a new client whom I probably won't see. Most of my visitors are young people with no work, coming to pay their respects to an Elder who might also "give them a break" one day. These are the "Unemployed Graduates" who plague every African government (every European and Arab government as well, come to that): educated for jobs that are not there. It is a particular problem in francophone Africa, where education is even more out-dated than in the primary schools left behind by the British colonial system.

Now I do not want to be completely unfair: the French education system in Africa has changed a bit since the nineteenth century. After Independence, for example, the Africans dropped the classic history textbook used during the 1940s and 1950s which began: "Nos ancetres les Gaulois....." When you discover that the first black political leaders of independent Gabon and Cameroun and Ivory Coast and Senegal were forced to learn off by heart about "their" ancestors the Gauls (not forgetting their Uncle Charles de Gaulle), it is easier to understand the love-hate inter-relationships which still exist between France and its colonies. Oops, sorry! I meant to write "Ex-Colonies."

The unemployed graduates are the victims of this colonial education system, which was conceived for one purpose: to produce secretarial staff for the colonial government. Schools served a paternalistic and centralising government machinery which, in most of the countries of francophone Africa, has bankrupted the nation and grown fat at the expense of a despised, exploited peasantry.

Salif, who had very kindly "called to see me," is a victim of this still-colonial education system. He handed me a letter from his Big Brother Douba in Ségou, former capital of the Bambara Kingdom. Ségou is where the Scottish doctor-explorer Mungo Park, in 1795, became the first European to see the Niger River. Poor old Mungo Park couldn't visit the city, because the king wouldn't trust a White Man to cross the river. History proves that the king was wise to be suspicious.

Douba is one of my oldest friends. If Douba has sent Salif to me, I cannot turn him away (worse luck!). "My small brother Salif is in need of assistance. He will tell you his story and I am certain that I can count on you to give him good advice" wrote Douba. For advice, read money.

Douba has seven children and a good job, which brings in a salary barely adequate to feed them all. As a loyal son, he sends money to his parents, and also to the parents of his lovely wife Safi. Rigidly honest, Douba has to make do with what the government pays him. He certainly hasn't got enough to fund indigent brothers. So, he honours me as a brother, by sending Salif to my office. I put aside my quarterly financial report and listen to Salif's story.

Salif trained as a primary school teacher. His wife is pregnant with their second child. Because he cannot feed her, she has gone back to her parents' house. Salif retired from teaching at the age of thirty, taking advantage of $6000 in cash funded by the IMF and World Bank. This is how the World Bank forces governments to reduce the numbers of public servants on the pay-roll: by losing teachers.

To support the recycling of retired civil servants, the European Commission is offering credit at 8% interest to help people start new businesses. Salif reckoned that with his capital of $6000, he could get a loan at 8% and start a business. He wanted to run a traveling pharmacy for the twenty villages near his home. His father is a Village Chief and a very well-known magic charm seller, so Salif would be accepted in all the villages (and he might very well make more money from the para-medical side of his work, than from the pharmaceutical). Salif presented his project to the European Commission. It was refused. This is a typical case of "cold money" institutions being unable to adapt to the needs of the rural population. The entrepreneur brought his own risk capital, yet the bank still refused to back the project.

Salif was stuck. He didn't have enough to launch the business with his $6000, because he needed to buy a van for the traveling pharmacy (I had to agree with Salif here: you cannot travel a pharmacy without transport). So, he changed to a sedentary enterprise, moved back to Ségou town, and installed himself as a grocer. He had just had time to purchase his stock and become known by the neighbours as a friendly fellow, selling good quality groundnut oil, when the popular revolution hit us all in March 1991. The military government was overthrown, general rejoicing and looting took place, and Salif's store was ransacked in the process.

"The rains have just started, Salif, so why don't you go back to the village to plant millet and maize, while you decide on the next step?"

Salif said it was difficult to go back to the village. At each prodding, he came up with another difficulty. His reasons sounded bogus. There were too many brothers. He could not go back without taking supplies or money. His wife was with her father. His plot of land was too far from the village to work it without living there. He would build a house first, and till the soil later, but you cannot build a mud house during the rains. His land was good for trees, more than for millet. He wanted to irrigate the land so he needed a motor pump first before returning to the land. In short, Salif didn't want to go to the village. What he wanted, he thought, was to start up again as a grocer. We did some calculations about margins for tea and sugar and oil. I told him to think it over and return next week.

Next week (April 1991) Salif had a different plan. "I am decided to go try my chance in Liberia. There the country is more peaceful now. The ceasefire has been announced on the radio this week. I can make money if I go to work in the reconstruction. I wish to save money for a grain mill and a motor pump and then I can return to the village."

I said I thought that Liberia might have too many bullets. "This is no problem. I know that no bullet can enter my body. My father has been attacked during the Burkina-Mali war with guns and grenades and mortars and no one could hurt him. So, I am not afraid of bullets."

Salif is a Bobo, and his protective charm is based on a cotton thread spun by a girl who has never made love with a man, with knots in the thread made with saliva and incantations from the fetisher, the whole being manufactured on the same day (which must be a Friday). This grisgris works better, Salif tells me, than his shirt covered with individual jujus, which can stop light bullets but apparently cannot stop heavy bullets. I think am prepared to accept this bullet stuff, since I have no basis on which to challenge the beliefs of Salif and his culture. It is a low-risk business for the fetisher, since disappointed customers are unlikely to return to complain that they were killed. The Western "scientific method" and so-called rational thought has such strict spiritual limitations, that they dismisses anything they cannot understand. I do not pour scorn on ideas that I do not understand. Perhaps Salif really is impervious to bullets. In any case, I am not going to test it myself.

I asked, had he discussed his Liberian plans with Uncle Douba? "No, I have not discussed this with Douba because he would not agree to me going to Liberia." I told Salif I would have to speak to Douba: he came to me from his-brother-who-is-my-brother, and I could not take the responsibility to send him Liberia without consultation. I dialed Ségou, greeted Douba in warm fashion, and turned him over to Salif for a discussion in Bwa, the language of the Bobo people. After 6 minutes it was my turn.

"My brother Lacville, I have to tell you that I do not agree with this Liberian adventure. But we shall have to let him go to Liberia." I was surprised. "Well," Douba continued, I w"ould not want to go to Liberia at this time. It is true what Salif says that the bullets cannot enter his body. That is a fact, but it will not protect him against somebody with a hammer who may hit him on the head. Liberia is, in my opinion, a dangerous place. The main thing is that he tells me he has some friends there in Monrovia. If they are still alive, then he may find them, and they will help him. Liberia is dangerous, but Salif is a young man and he is strong, and if he lives, he may make his fortune. Each man must seek his own destiny."

I protested for a last time that instead of leaving his wife and kids, Salif should be producing food to feed them. "That is true, Lacville. I would prefer that Salif should go to the village. His father is rich, there is plenty of land, and Salif can share the family wealth provided that he contributes to it. But the problem is that he does not want to work the land. If Salif returns to his father's house, he will be sent out to plow the fields. But Salif has been to school and he will do anything in order not to till the soil."

Douba sighed on the other end of the phone line. "You know, Lacville, we have made our education system so that even a boy who goes to school for one year or two years, he then despises the work of farming. After two years in primary school, he feels he is an intellectual. Now I have spoken with Salif, Lacville, I know that he will prefer to clean out gutters or to carry bricks on a building site rather than till the soil on his family farm. If we stop Salif going to Liberia, he will stay in the city and do nothing. In that case it is better that we allow him to travel to Liberia. If he is successful there, maybe he can gain enough money to buy the motor pump and then he can return to the village feeling proud. But I am afraid he will never accept to be a farmer again because he has been to school."

At Douba's request, I gave Salif some money and he disappeared. I met Salif again a couple of years later in Ségou, and I asked him how he had fared. "I was able to purchase an old Landrover motor in Guinea, and brought it back to Mali and sold it. This was good."

CHINESE MOSQUITO COILS AND POVERTY

Reading my Guardian Weekly, swaying in a hammock beneath the great mango tree, I half hear a shuffle. "Roberts, ça va? How is your health?" I look up unwillingly. I know the shuffle. I know the voice. It is Nafo. I seldom see him now, and he only comes when he needs money. He gives a nervous laugh. "Roberts, how are you? How is your health? How is the health of the people of your household? How are the children?" He giggles, and looks down at his plastic flip-flops. I notice that one of them is torn, tied up with a piece of plastic bag.

I can smell Nafo at ten yards. The stench of sweat mixes wretchedly with the stale smell of cigarettes. There is another odour as well: the smell of poverty and failure. Nafo is the father of Nana and Ousman Robert and Baby Jeanne, and no one looking at Awa sideways would be in any doubt that he will soon be the father of a fourth child. The pregnancy seems to be about his only achievement this year, for Nafo is not a very successful father economically. He manages to get occasional gardening jobs; but these day wages are not enough to make a living, and his wife Awa has thrown him out of her life.

Nafo has real gardening skills, and he sells plants down by the river near Bamako's only bridge. But the plant market is not huge and Nafo has a lot of competitors, many of them more dynamic and more educated (and more intelligent and better businessmen) than he is. Sales are slow, especially during the rainy season when so many wealthy Malians and the foreigners are away on vacation.

The economic crisis has bitten deep into the earnings of plant sellers. People must feed their children today, before they spend money on a mango seedling that may not produce its first fruit until it has been in the ground for seven years. In these times of political change and economic risk, who apart from the French and the Canadians can focus their time and money on buying flowering shrubs?

"If you will give me 7500 Fcfa, I will buy one carton of Chinese mosquito coils. Then I can sell them and I can make 2500 francs profit." Lying in my hammock, I ponder the matter. I remember all of Nafo's failures with the left side of my brain. But with the right side I feel guilty about my own easy life, my range of opportunities, my luck as compared with Nafo's hopeless struggle for survival. Nafo is a victim of outmigration. He arrived in Bamako full of hope but with almost no skills that would help him to survive in the city. He was a gardener with ambitions to become an urban businessman. He tried selling cigarettes, but he smoked most of the profits. Technically, they were my cigarettes. How many times must I fund his misguided conviction that he can get rich through commerce?

I remember the time he deserted Awa and her three children and fled to the Coast. I remember the photograph he sent her six months later. It enclosed a bank-note of 5000 Fcfa. After six months, he had finally sent these miserable $15 that arrived at Jeanne's office address. I remember Jeanne handing the photograph to Awa. She stared at the smarmy smile, the new clothes, and the new transistor radio in the photo. Then she picked up the 5000 Fcfa, and spat.

When Nafo was our gardener, he lived with his kids inside our compound. Then he quit gardening to become an unsuccessful trader – in the belief that commerce would make him rich. There are lots of rich merchants, of course: but they have skills that Nafo lacks. When we heard that Nafo had vanished from Bamako, we drove to Banconi in our modest-yet-luxurious car to seek out his family – our family – in the Bamako slums where he had dumped them. Nana was sick, Ousman was hiding, Baby Jeanne was covered with sores. We brought them home, fed them, clothed them, and sent the children to school. Nana was no longer ill. Ousman was smiling and cheerful. Baby Jeanne became fat and happy and beautiful.

When Nafo and Awa first came to work with us, Nana was a tiny baby on Awa's back and Nafo an excellent gardener. When we moved to The Gambia, Nafo said he would come with us. And he did. We all lived beside the Atlantic Ocean. Nafo was still our watchman and gardener, but seaside gardening is pretty meager. He mostly grew mango and pawpaw saplings for me to plant in rural school gardens up-country: he was good at this, but it did not take much of his time.

Nafo wanted to try his hand at commerce. I bought him a wooden table and a stock of cigarettes, matches, sweets and stuff that he could sell at the side of the road, in front of our front door. Nafo's stall became a social focal point for Malian exiles. Awa braided women's hair under a mango tree beside the market stall. Their daily life was pleasant. They gained prestige as Malian exiles. Slowly however, Nafo realized that he was smoking his profits, and accepting to give credits that ate away at his working capital and were never paid. We had been together for nine years and he still had a gardener's salary, but little by little Nafo was failing in business. He was becoming a disappointed man.

We returned to Mali, and Jeanne gave Awa money– mostly unearned - to keep her children healthy. We said it was for helping at the house. I wanted to set Nafo up as a fruit-farmer, to buy him land and sink a well …. but he refused: Nafo wanted to be a businessman! Nafo saw working the land as socially below being a merchant. I praised him as a skilled horticulturalist, and said I would not throw away any more money on funding his failed commercial dreams. He resented the money we were giving Awa for the children: so Nafo took his family away in a fit of spite. He moved them into the Banconi compound of his elder brother, who already had four wives and fifteen starving children. Instead of caring for his children, Nafo dumped them on his poverty-stricken brother and disappeared. We retrieved Awa and the children. Nafo became a memory. Then he became that photo, the one that made Awa spit.

Finally Nafo returned to Bamako, bringing with him the rows and dissensions that haunt so many families who live below the level of subsistence... where the struggle for life leaves no time for love. The last time I had seen Nafo, Nafo had been taking drugs. There was a row between Nafo and his wife. Awa proved to have the stronger personality, and a far richer vocabulary of Bambara insults than her man. He slunk away in defeat. Now he has come back to haunt me, removing my one hour of relaxation in my hammock.

Nafo shuffles awkwardly at the far end of my hammock. My Guardian Weekly has slipped off me onto the soil. I pick it up with a curse, blaming Nafo under my breath in a paroxysm of injustice, which can only derive from my feeling of guilt. I feel, sadly, that I have failed Nafo. Nafo has also failed me. I wish Nafo wasn't here. Nafo's shoulders sag these days, a telling sign of defeat. I remember all the insults he had heaped upon my head one day, probably under the influence of drugs. He even brought Ousman Robert along once to repeat like a little black parrot, a series of insults about my father's genitals. Those are the worse Bambara insults you can spit out, whether you are drugged or not. Ousman Robert was four years old at the time.

I groan inwardly at the injustice of urban migration. Here is a man who was a fine farmer. His father was the distinguished Head of the Poro Society in his Senoufo village, near Sikasso. I have seen Nafo grow champion maize crops. I have seen him caress mango trees with love and skill – at least 1000 of his mango seedlings are now growing in school gardens around The Gambia. Those skills are useless in an urban slum. I once bought him a pair of oxen and a plow to help him start up again in the village. But Nafo is too proud to return to the village. He gave them to his father. Since the old man is 90, this means some elder brothers received them. I suppose it helped the extended family and improved Nafo's personal standing, but it probably also added to the illusion of success and encouraged some other young villager to leave the land to try his luck, following Nafo into the Big City.

The wretched Nafo is still there, damn him. Why doesn't he go away? He only needs $20. "Listen to me Roberts: I can sell a carton of mosquito coils in only one day, so then I can give you back your money." That sounds pretty unlikely, Nafo. I have known you for years, and year after year I have seen you fail as a businessman. If you can make a profit of 2500Fcfa in one day, when a daily labourer earns around 700 Fcfa ($2), then everyone would be selling coils and the profits would fall at once. But I don't say that. Perhaps he can sell them in two or three days? Why am I even wondering about this, when there is no question of repayment. Even if he came to repay me, I would tell him to reinvest it in another pack of ineffective coils until he stopped the loan fiction himself. There is no lending culture in Mali: if you have it, you share it.

"Here are 7500 Fcfa. You can repay me 500 Fcfa each day. And here are 200F for soap. Go and wash. You stink!" Now that Nafo's not holding his breath, I hold mine – literally - as Nafo comes round the mango tree to collect the money. "Dis is very good, Roberts. Dis is a help between us. Dis time I shall be successful, surely very successful." I pick up my Guardian Weekly again. I shall not see the money again, but at least my conscience feels easier.

Nafo is still there. I lower the paper resignedly. Nafo hangs his head. There was a time when he didn't need to hang his head, when he was a proud gardener, full of optimism and fresh home from the village. Now he is beaten. Drugged by the need for city life as much by his cigarettes and cola nuts and occasional hashish. "It is about Ousman Robert. It is time he should be circumcised."

I remember instantly the day that Nana was excised, five years ago. I had been out in the bush for a week. When I came back, it was done. Jeanne had paid out the money to have a clitorectomy done in the hospital, and I was furious. Jeanne and her organisation fight genital mutilation; I loath excision and everything it symbolizes. "But you cannot fight against tradition, Robert," sighed Jeanne very reasonably. "At least they don't go in for infibulation."

Now it is my turn to provide the money for a dangerous ritual.

"There are two blacksmiths in the village," Nafo offers. I react without even taking the time to think. Using Jeanne's strategy, I blurt out: "You cannot use the local blacksmiths: you must have it done in a hospital." Now I am involved up to my neck. The left side of my brain rebels: Am I even sensible? Blacksmiths have been carrying out circumcisions for a thousand years. Why change the tradition? Blacksmiths must earn a living too. The right side of my brain replies: "Yes, and for a thousand years small boys have been dying from tetanus, because the blacksmiths knew nothing about hygiene."

I once spent the night in a Mandinka village in Senegambia, where the *Kampo* spirit was rampaging through the forest searching for evil spirits that had infected the penis of one small boy among the dozen who had lost their foreskins the previous day. The circumcising blacksmith may very well have been the man under the *Kampo* spirit mask, clashing his cutlasses and terrifying the whole community. Even the dogs and monkeys were silent; the very forest held its breath as the avenging *Kampo* spirit searched and stomped. The spirits are all-powerful when they are angry. If the *Kampo* spirit meets and kills a villager, it is not considered a crime under Gambian law. It was a very dramatic evening, as we held our breaths and cowered in the small mud house where I was staying with the head teacher of the school. The *Kampo* with his clashing cutlasses was throwing blame for the infection on spirits conjured up by the co-wife of the boy's mother. They always blame a woman.

"I cannot let my child..." The right side of my brain has won the argument: I shall have to fund a circumcision party at the hospital at the end of the rains. Nafo will never be able to afford it with the profits of one carton of Chinese mosquito coils.

A COY MOTHER TO BE

Awa is looking great these days. "Great with child," I mean. She already has three but no one walking round her these days could doubt that No 4 is imminent. One is not allowed to ask. I believe she would not dream of admitting it even to her husband. A husband gets to notice like everybody else, and the spouses get teased in about equal doses. You would have to be deaf not to hear the teasing around our household. "Got something to hide, have you?" asks the driver. Awa leans back and laughs. Well, does it look as if she is hiding it? But she'll never admit that she has a baby. "No, there is nothing" she retorts with a grin. "You have had a big lunch" I chortle. "No, I have had nothing to eat all day" she replies with a delightful pout. Then: "Feed me!" Her eyes sparkle provocatively beneath long lashes. Awa is one of those beautiful young women who look even better when they are pregnant. Awa glows.

Chet and Leïla aren't calling her "Awa" any more. Leo the gardener encourages them relentlessly to shout "fulana muso." This means "Mother of twins" (and not Fulani woman, which would be "fulani muso"). Not that they need much encouragement. When Chet is in good form, his 14-year-old voice croaks "Fulana muso, fulana muso, fulana muso" rhythmically to a hand clap while Awa contorts herself with laughter and shuffles around with the steam iron in time to his rhythm. Or she puts down the iron and claps her hands in unison while denying all the time that there is anything inside her at all! All very strange!

Then Leo the gardener fetches a stick from the garden and pretends to beat her, while Awa screams louder with her laughing. Or if she is unwary, Leo will catch her with a deep toe-kick from behind, suggestively low between the buttocks. In this case, it is Leo who roars with laughter while Awa chases him furiously into the garden. It is all very suggestive. After all, Awa and Nafo have not been getting on too well lately. Old Brother (who is a feudal family patriarch as well as an old gossip) mutters that there is a certain Amadou whom he fired, but who has been seen prowling around Awa. And Old Brother grunts that he wouldn't trust Leo. In view of Nafo's declining morale, and the signs that he is losing the struggle with urban living, I admit that I sometimes wonder myself who is the father of the child - or could it really be twins?

[Note: I suspected that Leo was the father, especially after the baby was born and Leo took such a great interest in the little boy. The truth is that this baby became a man who is now - in 2024 - a gangly 31-year-old, very tall and he looks like Nafo. Leo was a foot shorter than Nafo. Leo was probably more interested in the mother than in the baby, but Awa was not interested in Leo.]

Supposing it is twins? What would be the consequences? Sure, twins in the Sahel are supposed to bring luck (although in Africa's forest zones they are often seen as bad luck: how can a woman living in tough circumstances expect to feed two babies?). Let's think about it. Instead of four children, Awa would suddenly have five children in 1991 and how lucky is that for someone with few resources? Awa already has Nana (aged 12), Ousman Robert (9) and Baby Jeanne (named after my wife when she was born in our house in 1986). Awa's family has no income except what we are giving them, and occasional selling or gardening income from Nafo. Awa's children are dependent on Jeanne and me. Twins would mean greater risk, rather than greater luck.

Our American friend Carol (who works in family planning) comes to visit. When she hears the discussion about Awa's new baby, Carol asks a question which comes naturally to her, but which I never thought to ask. "Why is Awa still having babies?" I confess that I have no idea why. There again: Why not?

"How come you don't give Awa access to contraception and give her a chance to manage her life?" cries Carol in the voice of an American feminist filled with certainty. I look down at my dust-covered shoes, and kick a fallen twig of eucalyptus. It is perfectly true that I have never asked myself the question; nor have I asked it of Nafo or Awa. Africans seldom talk about sex, although they enjoy doing it. I have no answer for Condom Carol, but she hasn't finished with me yet!

"You're completely irresponsible: making Awa more and more dependent, less and less capable of looking after her offspring." Jeanne comes out into the garden. She will help me to defend myself, but she also doubles Carol's audience. "Listen, you guys" Carol expostulates, "If Awa had just two children, she could go off, leave her jerk of a husband, and be independent. But how the hell can she do that with four—or five? How come you don't give her access to contraception? You are not giving her a chance!"

I wonder where Carol thinks Awa would go, dragging her children behind her? Two children or four, the reality is that Awa will not go anywhere. We are her sole source of income, her only social safety net, and probably her best and most loyal friends in Mali.

On the other hand, I suppose Carol is right: Awa has a right to contraception. But should I be blamed for not fitting an IUD inside Awa? Is that my job? Jeanne thinks the birth spacing probably shows that Awa is planning her family, but she admits that she hasn't actually asked her. Our gynecologist friend Fanta is an expert in contraception, including tying fallopian tubes. Fanta speaks the right language both linguistically and in terms of being able to talk to shy women from the villages. We will ask her to talk with Awa.

Awa was born in a remote and very poor Bambara village; she is a product of urban migration, and she has become an urban woman. Why wouldn't she take "precautions"? Well, it would be because she doesn't have the information. Very few African countries have had an active family planning policy. Children after all, are a blessing, a source of prestige, and an economic necessity - the only source of security in old age. As long as nearly 50% of babies die before the age of five, African couples need to produce plenty of kids as an insurance policy. While the International Planned Parenthood Federation has nobly supported and funded many local Family Planning Associations, these have so far had little statistical impact on birth rates in urban West Africa, and none at all in the villages.

Even if smiling, lovely but illiterate Awa does somehow have information about contraception, there are other hurdles to overcome. Awa cannot afford the pill. Even if the pill were free, she couldn't afford the medical tests required to get a prescription. And legally she cannot obtain the pill without her husband's consent. If she had it, Awa would probably share her pills with neighbours complaining of a headache or a fever. The splendid African tradition of "sharing" can have disastrous results on the contraceptive cycle. Diaphragms? Hygiene? Creams? All costly and complicated. The best bet would probably be an IUD or an injection: the latter is becoming one of Africa's most popular contraceptive methods, because it is the only one that the husband cannot see. The contraceptive injection gives women the option to decide on pregnancy.

And what about condoms? My guess is that Nafo would never agree. Few Malian men would agree to use a condom with their wife. Even many sex workers do not use condoms. This is a culture where prostitutes charge 500 Fcfa with, or 1,000 Fcfa without – even though the girls know perfectly well about the risk of AIDS. The condom is a difficult "sell" in a culture where a woman must not even see the man's organ. The Malian government is actually listening to US government advice about promoting condom use, and there is now an AIDS awareness campaign. Condoms are coming in slowly thanks to AIDS, at least in the capital city; but at the moment most women are too shy of their partners to take a lead in contraception.

They are not shy of pregnancy, however. The reason for Awa's coyness about the imminent baby is fear of bad luck. If you admit to carrying a child, you may attract the attention of evil spirits. Far better leave the spirits in ignorance, and keep your child protected. The same reasoning explains why you should never congratulate an African mother on the beauty of her child. You can give her a gift of money to buy soap or baby clothes, but you must not attract bad luck by saying the baby is beautiful. Come to think of it, that is a sensible rule for other reasons. After all, what mother doesn't think that her baby is beautiful? And if she is already convinced, why waste energy repeating the obvious? If we could install this as a universal law, we could save an awful lot of hideous cooing from aunties and grandmothers.

Now our whole family is waiting with bated breath for Awa's baby. Could it really be twins? Will this baby be born as dramatically as the last one, which popped out unexpectedly in the early morning hours and almost turned me into a midwife? More prosaically, will it be a boy or a girl? And since the last two were named Ousman Robert and Baby Jeanne—whose name will be used this time?

AN AFRICAN FAMILY BIRTH

It happened while I was at the office. My two elder children were due to go to their piano lesson with a Russian lady, and they were piling into the car with sundry homework tasks or (more likely) illustrated Tintin books to occupy the afternoon while waiting for their lessons. It was then that Awa told Doulaye the driver that she needed to be dropped off at her mother's at the other end of town.

Naturally she wasn't talking about her mother at all. Her natural mother lives in a small and remote village some 120 miles away - which hardly counts as "the other end" of town, whatever one may think of the unbridled urbanization which is sweeping Africa at a greater speed even than "democratization" or AIDS.

Awa had been looking very pregnant for rather more than eight months, and she had become weary of the many jokes about being the "mother of twins." The time for joking had passed. Right now, Awa felt the contractions coming on, and she needed to go off to the home of an elderly lady from her village who would minister unto her feminine and maternal needs (including pushing down on the uterus, supporting her back while squatting, muttering appropriate feminine magic incantations, calling in the midwife to cut the umbilical cord, and dealing with other jobs which can normally be handled only by post-menopausal women who are "family" members). The elderly lady was to act as her "mother" by right of age and by right of village familial linkages.

I will not deny that the roads to our house in Magnanbougou are pretty bumpy, but I do not think they are entirely to blame for the fact that Awa was groaning and panting within two kilometers of leaving the house. Before Doulaye even reached the tarmac road, Awa had changed her mind about the other side of town and was asking for the nearest maternity clinic. Doulaye is not yet married. He is a respectable young man with no desire whatever to get mixed up in women's business, yet the poor fellow was driving a car with groaning Awa, a lot nearer to an active birthing that he had been since his own mother had brought him into the world about twenty-nine years ago. Gritting his teeth, Doulaye swung the car around in the opposite direction, away from the Russian piano teacher, and sped up the tarmac road. His speed increased as the groaning intensified. This frantic dash culminated with Doulaye swinging in through the gates of the maternity clinic at precisely the second when Awa's groaning turned into a shriek. Doulaye braked to a halt, keeping his eyes firmly to the front, and his hand firmly on the horn. Nurses and midwives came running, the baby was delivered on the back seat of our Peugeot. Mother and child were then helped into the clinic. "And what a mess she made of my car!" wailed Doulaye later when he told us about the event.

we heard most about it from eleven-year-old Leïla. "It was quite disgusting!" she declared firmly. "There was a lot of blood and then Awa did other things too. I was in the front seat with Chet and I decided to look out of the window and not look round at Awa at all after that."   We told Leïla she was very lucky to have been privileged to witness such an important event. When Awa's last baby was born, I was woken by Nafo and I found Awa at 4.30 in the morning, sitting naked on the floor of her hut with a baby girl between her feet.... while her husband Nafo kept well out of the way, as befits a husband. It was Jeanne and our neighbour Dr. Katy who did the honours of cutting the cord, checking out the placenta, and washing the baby (while I boiled a kettle and tried to sound heroic from the safety of the kitchen).

We argued with Leïla that she had had the best of the birth: all the pleasure, none of the pain, and none of the clearing up to do afterwards. And being good Africans, we congratulated Leïla on being witness to the birth of a boy. She refused to be convinced. We told Doulaye to take off the seat covers and to get a woman to wash them since they had been soiled during women's business. That made him feel better. So, I do not believe that any one had anything to complain about really. Chet did not complain at all. No wisecracking from him this time. He kept his fourteen-year-old tongue silent and spent a long time digesting the experience.

At 6 o'clock in the evening we drove as a family to see the baby boy. Although Chet was still keeping his counsel, Leïla had recovered and was looking forward to seeing her new plaything. Jimmy was jumping up and down on the back seat piping in his shrieky voice "What shall we do with Awa's baby?" to the tune of the "Drunken sailor". (Since Jimmy would not have thought of this by himself, presumably big brother Chet had not been silent all afternoon.)

"How long would Awa spend in the clinic?" asked Leïla, getting more and more excited at the prospect of the new baby. "Not long," I said wisely. "But she will probably stay a couple of days to get rested and to see the baby is fine." I was quite wrong.
There in the courtyard were Awa, together with husband Nafo and her children Nana, Ousman Robin, and Baby Jeanne (no longer the baby now). And quite clearly, Awa was waiting for us to come and drive her home only 4 hours after the birth. 2,500 Fcfa ($10) were needed for the clinic. I handed the money to Nafo and he took it to the midwife. Then we thanked everybody in a white overall, shook hands with the numerous people sitting around in the courtyard, and drove Awa home. Awa looked quite slim when she started carrying the washing basket out of the house next morning. I told her off for being silly and Jeanne ordered her to take a full week's rest before doing any work. Awa seemed pleased, if surprised. She was brought up in the village, where women never rest.

One week later I paid the money needed for a naming ceremony. Nafo told us that the baptism would take place at 6 a.m. When I warned Chet that I would wake him early, at 5.30 instead of 6.00. He grimaced and asked why he needed to come. I told him he was old enough now to start taking on his social responsibilities. But I also had a hunch that he ought to be there. As the sun began to rise through the pawpaw trees and across the Niger River, men began to walk in through the gate. Most of them were dressed in shabby work clothes. This was no affluent ceremony. This was a baptism for urban migrants surviving in the city slums. The elders were seated on their woven mat around the imam. Another twenty men or so sat on chairs and benches. Someone began handing around the cola nuts. I took mine and told Chet to take his. "What do I do with it?" he asked in amazement. "You don't chew it, that's for sure: it is disgustingly bitter and it is also a drug. It's a strong stimulant. It's what gets the American middle classes hooked on Coca-Cola. After the ceremony, you'll take yours and mine and offer them to Selima's grandmother who will be delighted both at the honour of receiving cola nuts and also because you brought them to her. She has no teeth and cannot chew them, but she will pass them on to some of the labourers, and they'll be pleased too and will chew them quite happily."

Everyone seemed to have arrived. Prayers were said. We all raised our palms to the lighting sky while incantations were muttered in an archaic Arabic incomprehensible to anyone (probably including the imam). Someone spoke the word "Chet." My son looked at me in surprise. "What is the baby's name?" he asked. Nafo came up beaming: "Now we have a Chet, as well as a Robin and a Jeanne. Soon we shall have a complete African family for you." Chet was thrilled. I was horrified. Carol is right, I must get Ami onto family planning ASAP.

WOULD YOU AGREE TO BECOME A SECOND WIFE?

Leïla is going into jewelry in a big way. See her in the distance and you'd think she's nearly ripe for marriage, with her tight short skirts and her dangling earrings. In fact, she is not ready for marriage (and I am her FATHER so you had better believe it). Leïla is nowhere nearly ready for marriage. Even though she may behave like an advanced teenager, she isn't quite twelve years old yet. But there is a non sequitur in the last two sentences. As I write them, my fingers are slowing up on the word processor.... for there are plenty of children aged twelve who are considered ripe for marriage.

In his amazing 1982 novel "Cobwebs" ("Toiles d'araignées" is the original French title), the author Ibrahima Ly recounts the tragic story of teenaged Mariama, born into a polygamous family and forcibly married (against her mother's wishes) to a wealthy old man in her father's Malian village. He rapes her. When she repudiates him publicly and refuses to marry him, Mariama is thrown into prison by the authorities and tortured – tortured and finally killed by the rules of the corrupt, male-dominated society into which she was born. I have just read in the press of an Indian gentleman who has been arrested for kidnapping an even younger girl. The child had been blubbing all the way across the Indian Ocean and a thoughtful airhostess rightly suspected that she had been abducted by the man in the aircraft seat beside her. On being arrested at the airport in Europe, the Indian gentleman protested angrily that this was no abduction. He had paid a high price for this child and he provided a marriage certificate to prove that she belonged to him.

Not a week goes by without some African man making a joke about marrying Leïla. She takes this perfectly calmly as though it were quite natural. Since she has been in Africa since she could walk, it is natural. This joking about marriage is part of

West African humour, just as polygamy is part of West African family life.

"How is my wife?" asks the bread seller when I call in for my evening order of two baguettes (French bread in a French colony, of course). When we stop to buy bread from the bakery, Leïla often offers: "I'll go and buy it, my husband is there," going along nonchalantly with the bread seller's joke. It must means that she likes him, thinking him harmless and amusing.

The watchman on my office also calls himself her "husband," and Leïla likes him too. However, this week he has had his head shaved clean to keep it cool for the hot season and to keep away the lice: with a bald pate, I reckon his chances with my daughter have gone down quite a bit. My Old Brother sometimes refers to his younger son as Leïla's "husband" as well. This is even more improbable than the bald watchman, since the son – 3 years younger than Leïla - is so very like his dominant father. It would be a bit like incestuously marrying a younger brother, but without any of the advantages.

Leïla's love of jewelry is indulged by her mother. But it is actively encouraged by my friend the African Princess, who shares my daughter's passion for rings, bracelets and pendants. Admittedly the Princess prefers real gold, while Leïla is perfectly content with kitsch. And thus it came about that Leïla was locked away one day in the Princess's room to discuss "women's business" while I watched nubile Ivoirian dancers wriggling their ample jeans-clad buttocks on the television in a frantic effort to keep their hips in rhythm with Alpha Blondy's jogging. Perhaps you have never noticed that, while many West Africans are fabulous dancers, many are not. The famous Ivorian singer Alpha Blondy has a jogging dance routine which is as inelegant as watching an elderly lady in a sports hall desperately trying to lose weight by running on the spot.

Leïla's "women's business" turned out to be rummaging through the Princess's jewelry collection, and coming out with a double handful of baubles for her ears, wrists and throat. Malians are so generous, especially to young girls. Leïla was very pleased with her booty. But the most interesting part of the "women's business" was Leïla's conversation with the Princess and her big sister, known as "Baby" or "BB." BB has been married before, but it didn't work out. During the summer holidays, while we were in Europe, BB got married again.

"Did you really get married while I was in France?" asked Leïla. BB smiled, and said that yes, she really did. "And do you love him?" BB smiled and said that yes, she really does. Then Leïla thought of something and asked impudently: "And when you were sitting there in a white dress in front of the Mayor, did your husband opt for monogamy or polygamy?"

This is always one of the most interesting moments in a wedding ceremony, and everyone strains their ears at that moment to hear the bridegroom's response. But Leïla was not expecting BB's reply: "There was not really any choice to be made, since I am the second wife." And both sisters burst into peals of laughter at the sight of Leïla's eyes widening until they became like saucers.

Polygamy is said to be going up in the towns (29% of urban marriages are polygamous, 28% of rural marriages, according to the demographers). This is probably because more people are coming to town, and they often take a second wife when they become established. A new urban wife is appropriate for the new urban lifestyle. Towns also contain more affluent men than the villages: and one of the signs of affluence is to have a second wife. On the other hand, the statistics show another stunning trend: whereas twenty years ago divorce meant that men were repudiating their wives, now the boot seems to be entirely on the other foot. In 1989 they say that 52.7% of Bamako city divorces were initiated by women. And for 1990 it was 77%. If that is true, it is really quite a revolution in urban (and polygamous) living.

Polygamy is something we know about, read about, hear about: but for Leïla it was a surprise that two of her best friends should be involved in it, in such an intimate fashion. She recounted all this in the car, on the way back for lunch. In fact, she could talk of nothing else. 15-year-old Chet was very interested. And the following week when the African Princess was in the car with us, he started an interrogation. Why wasn't she married? "Well, I suppose I have never met anybody whom I felt that I could marry." Had she never met anybody she wanted to marry? "Perhaps." Would she marry someone who had been married before? "I might accept to marry someone who has been divorced. But I do not believe I would marry someone who wants to divorce his wife to marry me. You know, Chet, in Africa we say that if a man divorces his wife in order to marry you, then he will just as easily divorce you later on, in order to marry someone else."

That gave Chet some food for thought, and we gained some small respite before he pursued the interrogation. Then: would she marry someone as second wife? There was a fractional pause, before the Princess replied: "I might if I loved him enough." Well, she might, I suppose, since her sister did. People's ideas change. But the Princess is a modern woman, and a professional woman. I happen to know that a very good friend of mine did ask her to marry him, and she refused him because she could not accept to be a second wife.

MY DECEASED BROTHER'S CHILDREN'S SCHOOL

Gaston is strapped for cash. He is heir to 6 children and he has to pay their school costs: 3 uniforms each, books and notebooks and pencils and erasers and set squares, as well as registration costs and "solidarity payments" to encourage the teachers.

Gaston Koné's brother died last dry season. He was called Nestor, a 42-year-old Bwa or Bobo farmer. The Bobo are a language group admired for their hard work and solid agricultural production. There were three Koné brothers with the same father and same mother: now the two remaining brothers have to carry the load for Nestor, their departed middle brother. Albert the eldest is a farmer like his brother: so, Albert and his wife are housing and feeding Gaston's family in their home village. If you pronounce them in the French way, these names sound somehow more probable, more African. They are definitely very African. Albert is a good Catholic, and he is thanking God that the rains were good this year. He has also sacrificed a chicken, in order to ask the Ancestors to send one more huge rain at the beginning of October, to ensure a bumper harvest. No African is foolish enough to place all their confidence in an imported religion: whether Muslim, Catholic or Protestant, these guys all maintain their respect for the Ancestors.

Gaston is an engineer in the city. To Gaston falls the task of providing clothing and schooling for the six children of his deceased older brother. Gaston recently got married, and his wife has produced a baby girl. It was rather a shock for Gaston to find he suddenly had six extra responsibilities, in addition to his new baby.

Of course, there are other brothers and sisters in the village. Dad had four wives, and there are more offspring than Gaston can remember. But Dad's second wife had just the three boys Albert, Nestor, and Gaston and they support each other.

Balima in Bambara means "children of the same mother" as well as "loyalty" and "trust": people of the same mother as you, they are the ones you can rely on. In a polygamous family, other wives' children are more often rivals than true friends. Co-wives naturally encourage loyalty between their biological children, to the exclusion of most others. *Fadenya* = "children of the same father" is a word synonymous with "rivalry" and "jealousy." Brotherly love in the traditional European sense centres around the mother's hut. The father commands the family compound, and he commands respect. Love comes from the mother. Jealousy comes from the co-wives (or "Little Mothers").

Gaston's deceased brother Nestor had a tough life. Farming is pretty hard anyway, although Gaston's family comes from a Bobo region on the Mali-Burkina borderlands, that usually has reliable rainfall providing adequate millet and guinea-corn (sorghum) harvests, as well as a good cotton cash crop. Even with world prices sliding over the past 25 years, cotton farmers have been among the most affluent in the Sahel. Not rich. I would call them poor; but they are affluent by local standards.

In the Sahel, "affluent" means not much debt, not much hunger, and no starvation. Back in the day (in the 1980s and 1990s) many of these cotton-and-millet farmers owned a bicycle and a transistor radio, some plowing equipment and an ox or two. Only in very bad drought years like 1983-84, did these Bwa farmers have to reduce family food consumption to one meal per day in order to survive the pre-harvest "hungry season." [30 years later they all own a mobile phone as well, though it may not contain enough credit to make an outgoing call. Very often I receive just two rings on my phone, meaning the caller has no credit and is hoping I will call them back.]

Nestor, the middle Koné brother, had real bad luck with his wives. Are you thinking that "maybe their bad luck was with him"? You should not speak ill of the departed. Those who have gone before us, are become Ancestors. Albert and Gaston speak to their brother when they visit the Ancestor House in the village, and pour millet beer libations on the family fetish. The Late Nestor protects the family, along with the other Ancestors (at least I hope he does) and so, of course, do the Catholic saints: but, unfortunately, none of them feeds or clothes the children.

His first wife bore Nestor two children, before she ran off with a truck driver. She came back to the village after a while.... but then she ran off again, and her husband lost patience. After the divorce, Gaston's deceased brother Nestor took a second wife - a lady from the next village who had recently returned from the Coast. She had been possessed of a devil, but it had left her in Ghana where her brother worked as a docker. And she had returned. They did well together, survived the bad drought years of the early 1970s, and made three children. The third child was born during the second terrible drought of 1983. The wife was already surrounded by four hungry children, and now she had added a suckling infant searching for more milk in her thin breasts. There was almost no food. Gloom spread across the Bobo plains of Mali and Burkina. Maybe this brought back the devil. The wife ran away eastward to Burkina, and they found her in the bottom of a well with her dead infant still tied onto her back.

Nestor took a third wife, who produced three children in six years, and then her husband died. A lot of African men die in their forties. Either Albert or Gaston should have inherited the widow; but they are Catholics. The White colonial religions are intolerant of African custom: adopting the traditional solution and taking Nestor's widow as their second wife, would have been contentious. The problem didn't arise since the widow ran off with an army sergeant, taking the youngest child with her. That is one less mouth to feed, one less child to clothe: but six of Nestor's kids remain in the village. Clothing isn't a problem because the climate is warm: you might even call it hot, since

10 months of the year we reach 100 degrees Fahrenheit every afternoon (and for four months it is often above 120 F). Children can wear one of anything (to look clothed), and it doesn't matter how ragged. But school uniforms have to be made from regulation cotton cloth costing 1,200 Fcfa ($4) per metre.

Gaston had a good job until last month. He was working for a White builder known as "Mayonnaise." But alas, the White man is a crook. He survived and made money in Mali thanks to the corrupt patronage of an army general who is now in prison (after the people's revolution of March 1991, all army generals were forcibly retired, and some were imprisoned).
 "Mayonnaise" stopped paying his staff when the old regime fell. After three months, Gaston asked for his back pay and he was fired with no pay. The poor guy: first his brother, then his brother's children, and now no job. And school is starting as well.... Family solidarity will come into play. That is the only way the Africans can survive the hard knocks of daily life on the edge of survival. I have sent a message through another "brother" (actually a third cousin) that I am good for a contribution of $100.

BIRTHING AND CIRCUMCISING THE ROBERTS

We have been visited by the latest of those unfortunate children who have been baptised in honour of their parents' White friends or employers or benefactors. You already know about Baby Jeanne and her brother Ousman Robert. I haven't ever mentioned the small Fulani boy who must be able to walk now and whose life will forever be blighted by the fact that his mother baptized him Poulton Jallow. [Author's note: this I have changed from the original article's name of Lacville Jallow].

If you ever meet this Jallow, you will know where he got his first name. A quick glance will confirm that there is no blood relationship (I am sure that this was the first assumption which leapt into your nasty mind.... if I am wrong then your mind is not nasty, and I apologize: I was addressing myself to your husband). This baby got his name because his mother felt that I had been kind to her younger brother in the matter of health, eye operations, education, and general support (which included a plow and a donkey). Naming her latest infant after me was a way of expressing thanks. I was touched by the gesture, even if I felt sorry for the child. He must have a nickname (everyone has a nickname). I was director of the charity ActionAid at that time, so perhaps his parents call him "Director."

My wife Jeanne is very popular and she has been responsible for lots of namings: kids will like her name. Others are not so lucky. I knew a strong Nigerian man called Hyacinth: not an easy name to carry into maturity. In deepest Mandinka country along the Gambia River, lies a village in which Professor Margaret Haswell of Oxford University first carried out agricultural studies back in the 1940s. She has been a regular visitor over the years and there are several generations of children named after her. All girls. There is a Haswell Darbo and there are two Haswell Bojangs that I know of. Anyone got a stranger name to offer than Haswell for a girl? Poor Africa!

This new baby visitor is called Robert. He is plump and cuddly and looks healthy enough, even though the mother says she has very little milk. There is a local proverb which says: "If there's no milk in your mother, you suckle your grandmother," but Baby Robert's grandmother hasn't been pregnant for years and her breasts are like dried prunes. This proverb is the equivalent of the French proverb "If you cannot catch blackbirds, you'll have make do with thrushes." Thrushes are less tasty, but they are easier to catch in a net. We all remember from our nursery-rhymes that blackbirds are very good to eat: at least they were before the farmers started spraying the countryside with poison. From memory I believe that a decent family pie requires about 24 blackbirds.

Baby Robert was born on August 10th and he was brought to see his namesake six weeks later with the blood still wet from his circumcision. It reminded me that Ousman Robert (who is seven) is due for circumcision this year. I bought a pair of red football shorts for Ousman to celebrate our new joint namesake. He liked them a lot. Appropriately they are coloured blood-red. Why put a seven-year-old through the pain of a severed foreskin, I wonder, when it could be done at birth?

The answer to this used to be that boys endured "ritual initiation into the age group system": an important part of village life. Groups of boys circumcised together form a bond for life, and each age group has a specific role in village organisation. But here in the city there are no group-organised rites of passage, no initiation into the village age groups. In the town all races meet and mix. Their common rites of passage are Primary School, Secondary School, Examinations, Football matches, dating girls and – a lot too often – creating an unwanted extra-marital pregnancy. These new rites of passage sweep away the differences between language groups and ethic traditions: everyone sits the examinations in the same place and on the same day……………..

When it comes to circumcision, I am not quite sure which ethnic groups do what when. It is all very complex, especially since these children called Robert have fathers and mothers from different ethnic backgrounds. Who can say when the mother's influence will predominate over the father's, or whether their cultural traditions have been swept aside by urban living? One friend of mine had eight boys circumcised together, to save money. A single circumcision fee, only one party for the whole family, one set of new clothes each, and finished with it. He borrowed a car, loaded into it all four of his own and four of his brothers' sons, aged from 3 to 12 years old, and had them all done on a Sunday morning in the hospital.

Having admired the baby, and Ousman Robert's red football shorts, we proceeded to offer some money to buy soap for the new baby's mother, and we handed over some of our nice baby clothes, which Jimmy hasn't worn for years. Chet remarked sourly that they are never likely to be worn by anyone else. It is true that African babies in the urban slums seldom have their clothes mended (or even washed). The buttons on the romper suit will never be replaced once Baby Robert's clumsy nurses have wrenched them off. Sewing is a skill few Bamako mothers practice. When Baby Jeanne's tee-shirt has a seam going, it goes. All the way.

I admired Baby Robert's bracelets. On the right wrist he had a little bead bracelet with a round leather disk sewn onto it. This is a Bambara charm against the evil eye, or to defend him against any curse brought by another woman. "Also, if a woman who is a witch grabs the baby's arm, then it will not be harmed and the arm will not come off." That sounds like a sensible precaution. At only 100 Fcfa (30 cents) it's well worth the money. On the left arm Robert has a twisted metal bracelet: a Fulani charm against fevers. In malaria-endemic countries, malaria is a killer for babies and any protection is better than none. I believe Professor Brian Greenwood, the world expert on this subject whom we knew well when we worked in The Gambia (he is now Sir Brian Greenwood), would put more faith in mosquito nets, preferably soaked in an insecticide. I try to get the children in my African family to sleep under mosquito nets, but my success is mixed. If I am there and nagging, they hang up the nets. Otherwise they seldom bother. With or without insecticide, the important thing is to sleep under a net in order to stop the insects from biting kids at night.

The leather thong around Baby Robert`s waist proved more difficult to analyze than the bracelets. The baby's father grinned helplessly with toothless gums and said this was the women's business. His wife's podgy face contorted as she tried to explain through an interpreter that this waist thong is a charm against black shit. No, apparently it is not diarrhoea.

EMERGENCY WARD TOURAY

AWA came shuffling in as usual on Saturday morning, baby Chet strapped to her back inside a brightly coloured cloth printed with green and blue butterflies. She deposited the baby on the couch, and stuffed the butterflies under his legs to catch anything that might come out. I was cutting a juicy grapefruit, and carefully gouging each segment, using the curved knife with serrated edge which my sister-in-law found in Germany. If there is a useful gadget that can be invented, you will find it in Germany. I also received a pointed grapefruit spoon, which also has a serrated edge around the point to help collect the juice from the sides of the fruit. I sank my teeth into a neatly separated segment, running my tongue gently over the jagged edge of the spoon. Juice ran deliciously down my throat. This is close to Heaven. I closed my eyes, and dreamed of the Garden of Eden. Eden, I believe, was on the banks of the Nile. Eden was Paradise, and Paradise is close to Heaven. Once again, my belief in God was reinforced by the taste of heavenly African fruit.

"Chet, il est malade." Awa is standing defiantly by my breakfast table, a basket of dirty washing on her hip. I come out of my reverie. Gulp of tea, go and look at the baby lying on his couch. He looks plump and healthy, pleasantly quiet. Could quietness betoken illness?

"What is wrong?" I ask. "His lungs. His breathing, it is not good." Hmmm... is this just maternal worry? Is there something really wrong? As I finish my grapefruit, I work out my strategy. I didn't intend going into town just now. How can I avoid twelve kilometers of road, a long wait for the doctor, big expenses.... Have I any Bactrim in the house? A quick check in the fridge reveals only adult 500 mg tablets, and they are not for a five-month- old baby. And antibiotics are unlikely to improve a baby's breathing. Supposing it is serious? Jeanne is away visiting village clinics in the far-distant bush. I'd look pretty damn stupid if she came home to find we had lost a child, and I hadn't even taken him to the doctor! I had better trust Awa's maternal instincts.

"Awa! Let's go! Leave the dirty washing, and we'll go to see the doctor." It takes me five minutes to dress while I finish off the contents of the teapot. Awa appears transformed in a fine "boubou": long dress over a matching "pagne" wrapped as a skirt, and with headscarf of the same material. The cloth is only a cotton print from Ivory Coast, which Jeanne bought her for the end-of-Ramadan festival last Spring: yet Awa manages to wear the colors of grey leaves on a dark blue background as though the cloth were made for a princess. The washerwoman of five minutes ago is now an African Queen.

The nurse at the French Medical Center is the wife of my friend Roland Sidibé, and that is where I drop Awa. Awa can wait her turn, while I do some work. I leave a note for Brigitte, give Awa money for the inevitable prescription, and off I go. Twenty minutes later, Brigitte phones. "Baby Chet is hardly breathing. We are sending him in the Medical Centre's car to collect you, and you must take him to Gabriel Touray Hospital for oxygen. And you had better go yourself: if the mother goes alone, she will have to sit in the waiting line for two hours and it may be too late."

There were forty mothers sitting in the Paediatric Unit at Bamako's main hospital. Some were suckling fragile infants. Other were cradling children who were beyond movement. I presented Brigitte's letter to a gorgeous twenty-five-year-old, whose white gown matched the bright smile in her wide copper-coloured face. I hoped she might be our doctor. "This letter is addressed to the Professor, and he will not be in today."
I changed my mind about the girl: she might be lovely out of a white coat, but clearly she was not the doctor I needed. I left Paediatric III, and crossed the corridor to Paediatric IV. Young Dr. Dembele received me with a weary sigh. "There is already one patient stretched out on the couch, and I am dealing with

this lady and her premature baby. I cannot do everything. Anyway, the Professor is not here."

I put the letter down on the desk, stepped back, and surveyed the situation. The ten-year-old on the couch looked like he was finished. The minute creature in the lady's arms couldn't weigh more than 750 grammes. It waved a pink hand. The bones of the hand stood out through the skin, looking more like the hand of a frog. Could this infant possibly live? In any case, that was not my problem. Darwinian instinct took over: whatever the needs of the other children, my priority was Baby Chet and his needs came first!

I stepped forward again. If the envelope was addressed to the Professor, and if the Professor was not there, the first thing was to remove the envelope. I tore open the envelope, waited until Dr. Dembele had finished writing out his order for an incubator for the neonate, and began to read the letter. "Signs of pneumonia, associated with bronchial infection. Breathing rapid and laboured. Absence of dilation of the nostrils. Temperature raised to 39.5 degrees......."

"Very well, M'sieur. Please give me the letter." Dr. Dembele was efficient and helpful. Within five minutes, young baby Chet was in a cot in the intensive care ward, his legs beating the air as the nurse held an oxygen mask over his face. Awa looked doubtfully through the window. I encouraged her: "You see? No more problem!"

We waited another half hour for the prescriptions. I had time to admire the beauty and the fortitude of the waiting mothers. Many were villagers, or recent immigrants to the town. You can tell from their bearing, from their clothes.... There was not one single woman in that hospital waiting line who did not look magnificent, everyone of them dressed in striking yellows and blues and reds and greens and blues. Their physical carriage, their silent determination, their inner strength and the sense of resignation on their faces gave these ladies an aura of nobility.

Strong women in colourful, beautiful clothes are the glory of West Africa. Three women sitting together wore autumnal colors: orange cloth with huge brown flowers, great orange flowers bigger than their faces, printed onto a fabric of dark green and russet. They looked serene, hiding their worries as they sat patiently with their sick infants in their laps.

How long would they have to wait? By African standards Bamako has plenty of doctors; but the numbers are pathetically few by the inflated standards of the West. Westerners complain all the time, having no idea how wealthy and privileged they are. These Malian women would wait all day, because they didn't have a confident male like me to push them to the front of the queue. They had no authoritative White face to get them priority over baby Chet. Many of these women have never been inside a hospital before. Some of them have perhaps never entered such a frightening place as a concrete public building, a building with stairs, a government building, a large and confusing place bustling with busy doctors and nurses wearing white coats. Some rural women may not even speak a language used by the doctors and nurses.

How long will they have to wait? How are their sick children going to survive? How many of them have already lost one or more children? How lucky we are - and yet we take our good fortune for granted. How lucky are Awa and Baby Chet, who happen to have found a family to provide them with security, and protect them from the misery of the urban slums.

"It was good that you found oxygen for Baby Chet," my driver Doulaye told me. "When my sister's baby needed oxygen, the hospital had run out of oxygen. The oxygen bottles were empty. Her baby died."

IN THE RICE FIELDS OF TIMBUKTU

The greatest way to travel is by river. I spent a part of the Fall of 1991 drifting along the river Niger in a wooden canoe, studying the rice harvest. My project contains an important agriculture and food security component, especially in North Mali where starvation is a threat every time the rains are dodgy. If you can secure a minimum harvest of rice, you can avoid famine. I was visiting the rice fields of Timbuktu, and the only way to get there at November harvest time is by canoe ("pirogue" is the Malian word for a canoe). Lucky me!

In June 1990, armed Tuareg rebels emerged from the Libyan desert to initiate a popular revolution against the 23-year-old military dictatorship of General Moussa Traoré: Moussa had fallen in March 1991, and we were in a year of political "Transition" towards democratic government. I was not allowed to travel by road because of security issues around Timbuktu. This forced me to hire a pirogue. There is no hardship in lying back against the hand-hewn planks of Ghanaian hardwood, reclining in comfort on cushions, watching white egrets circle over the rice fields while the boatman's assistant brews mint tea in the bottom of the boat for my drinking pleasure and that of my colleagues. River is better than road!

Now please don't start thinking that I am not an assiduous worker. When I was not assiduously discussing the harvest with my African colleagues, I was assiduously drinking green mint tea while assiduously searching through binoculars, hoping to spot a rare Goliath Heron around the next river bend. My Goliath harvest was disappointing (actually it was nil), but the rice crop is excellent this year. The rain-fed millet and guinea-corn harvests I can see on the river banks also seem good. Things are looking up for Timbuktu, unless the Libyan-fed Polisario-led Tuareg rebellion leads to civil war and the break-up of West Africa as we know it. In that case it will be the mercenaries who benefit from the guinea-corn harvest, while the local population hides in the sand dunes, too frightened to finish reaping in their fields.

When did a farmer in any country admit to having a good harvest? We tied up at a wooden jetty in the market village of Kirchamba, where the Village Chief admitted that "Our harvest has been passable. But you must know this: one harvest is not enough to help us recover from many bad harvests. This year we had perhaps 250mm of rain: but last year we had only 73mm, and we harvested almost nothing. After three bad years, a single good harvest is not enough. Everybody in our community has debts. You cannot survive in a bad year unless you borrow from the moneylenders. Here we say the first good harvest goes to pay your debts; the second goes to repay your social obligations to those who have helped you; and it is only the third good harvest which will bring something for you and your children."

The chief survival strategy in Timbuktu region is migration. As soon as the crops are in, young men leave for the Coast. Very few try to reach Europe. Most men head south to Bamako, and then travel on to Ivory Coast, or Ghana, or Nigeria. They find labouring jobs in the plantations or the seaport docks, or they ply the streets as tailors and hawkers, making just enough to live on and barely anything to send home. A straggle of these migrants gets as far as Gabon or Zaire where they try their luck in the diamond business.... or to Angola where they may be swept into the civil war and killed. A few head migrants north to Libya. Some of those who migrated northward during the drought years of the 1970s, joined the Libyan army. These are the men who have returned home to Mali to seek their fortune as mercenaries …. to take revenge against Moussa's military regime …. to threaten our destruction.

Most African migrants (maybe 90%) move around Africa. The recent Ivory Coast census shows that about one-third of the ten million inhabitants in that country are foreigners (mostly from Mali or Burkina Faso). As the climate gets tougher, the Sahel is losing its population. Many migrants never return home. A few die in the desert or drown while trying to cross to Europe. Like in Ireland after the potato famine, the most enterprising youth of the Sahel are moving out. This is a land of heat and hardship and sadness, where the sand dunes are moving inexorably against the people of North Mali.

"The situation was so bad this year," explains Aissa, "that the mayor of my village had to send out the Town Crier to plead with the merchants not to exploit our poverty. In June the Fulani women were coming into the market to buy millet, and prices had climbed from the post-harvest price of 50 Fcfa per kilo, to an all-time high of 1500 Fcfa per kilo. That means they were paying thirty times more for their millet than what they received when they sold it six months before. Of course, they had no choice. Because they are in debt, the merchants force peasant farmers to sell their crops immediately after harvest, when prices are at their lowest. Later they still have no choice because their children are starving: they have to buy back at the high prices, the same sacks of grain that they were forced to sell at a low price. The Fulani women were forced to sell their last gold bracelets and earrings to buy millet, and the mayor was begging the merchants not to destroy the final dignity of these Fulani women, not to take from them the last vestiges of their pride and culture. He asked the merchants to give more credit instead of taking the gold."

"Did the merchants take the gold jewelry or not?" I asked.

"Merchants are merchants," sighed Aissa bitterly. Aissa is a Songhoy, a salaried agronomist working to improve yields and restore seed quality: "The merchant money lenders accepted to keep the gold on credit, pawning it so that the women have a chance to buy it back. But how many of them will succeed to buy back their jewels when they have such heavy debts?"
As we drifted along in our pirogue, other canoes were moving through the green rice fields, filled with harvesters. We were in a watery world of floating rice, indigenous varieties that grow in

the Niger's flood waters on stalks that can be up to two or three meters long. You can only harvest this rice from a boat. You can also only harvest rice if the rains and floods come at the right time during the growing season. The first rains allow the rice seeds to germinate, ready for flooding as the river rises gradually from July through October. If the flood is too quick, the seedlings may drown. If the flood waters are too slow, the seedlings will bake and shrivel in the Saharan sun.

The only way to improve the rainfall, is by praying to Allah. But you can control the flood waters in order to increase food supply. Aissa and her team are getting the villagers to build earth dykes, helping with cement and metal bars to create solid gates where the river water enters the fields. With simple wooden planks slotted into the concrete gates, communities can then control how much water enters: allowing in just enough water to feed the rice seedlings, and not so much that they drown. The gates include a metal grille that stops fish entering the fields to feed on the rice seedlings. Several NGOs in North Mali are running such dyke programmes, increasing rice yields. It is an effective, small-scale food security programme.

Aissa works for CARE, an international US charity specialising in things like dykes. To get the dykes built, they pay labourers with food. "Food for Work" provides in-kind wages, supplied by the US government as part of their aid program. Around Timbuktu, where there has been a shortage of grain as well as of cash, bringing in American sorghum can be helpful. Not only does it add to the local food supplies: it also encourages some of the men to stay at home and earn food with CARE, instead of migrating to the Coast to earn cash. Some men prefer to smuggle cigarettes or wave Kalashnikovs around for a living (kidnapping can bring lucrative ransoms), instead of farming along the Niger River. Insecurity threatens the livelihoods of everyone: it discourages farming and trade, and this increases the risk of hunger for mothers and children.

Mothers and their babies are the primary victims of violence, of hunger and of malnutrition. In their desire to increase food security, we even find women working on the dykes. In this Songhoy culture where women usually do nothing but cook, and make themselves beautiful, some mothers have actually become farm labourers in order to feed their children.

Western feminists are in a quandary here, aren't they? We have all heard that African women are overworked, but these Songhoy women are traditionally not overworked: they do not fit the stereotype! It is only because of climate change and drought that these women are becoming hired labourers: we are now over-working them for their survival. Well, survival comes first. Then development. If the women are proud of their achievements as dyke-labourers, maybe we can help them organise nutritional and income-generating projects like vegetable gardens, without the men, so they can achieve for themselves better health, improved nutrition, clean water supply, and self-initiated incomes. Maybe they could also persuade some rebels to settle down and become family men: to stop killing people, take a wife and set to work building dykes for the rice fields.

GOLD AND DROUGHT AND SEED BANKS

AFTER thirty years of drought, Saheliens survive on debt. When June comes around and you have nothing left in the granary to feed the children, what else can you do but borrow grain? This is the "hungry season" before harvest time. First you go to your brothers, then you try your sisters, and naturally this includes all your neighbours. If they are all in food deficit, there is no one else to go to but the merchant or the money lender. This year the 1991 harvest has been good. Good for whom? Especially good for money lenders. It is the merchants and money lenders who have first call on the harvest.

First you must pay your debts. Provided you pay promptly, the kind and wealthy merchant will be willing to give you some more credit next June, during the next "hungry season" when you most need help. He will probably loan you back at 150 francs per kilo the very same sack of grain you sold him in January for 35 francs per kilo. The planting season is the hungry season, because everyone is hungry except the merchants. Farmers work hard in the fields, while their grain stores are empty. Who can keep their granary full, when the rains fail year after year? Who can harvest rice in the Interior Delta of the Niger River, if the rains are so light in the distant mountains of Guinea that the river flood is insufficient to inundate the Malian rice fields?

I remember back in 1988, I went to evaluate a village agricultural rice-field flood-control project along the Niger River. I found there was no agriculture in the village that year: there had been too little rain even to germinate the seeds. All the men had left the village to find labouring jobs elsewhere. There was no one left in the village but hungry children, and desperate mothers begging fish from passing canoes. Oh, and a few grandparents dying from starvation.

1991-92 has a different feel to it. People are cheerful. Maybe the drought cycle is ending. If Allah wills it. This year the water table has risen in the wells. The river is high, water is plentiful. During my river tour of Timbuktu's villages and markets, the water was lapping the feet of the village houses, and we could drift our pirogue up almost into the middle of the market square. During the dry season, the river may retreat several kilometers from the village. From February to July, the small weekly markets are parched, dusty with the Saharan winds. On market day people arrive with their cattle and donkeys, and the central square becomes shrouded in a dust fog kicked up by the feet of 300 animals. But in November and December, these parched villages are transformed into attractive riverside resorts, where cheerfully decorated trading canoes tie up each Monday evening, preparing for the Tuesday morning market.

As the last crops are brought in during this January 1992, everybody admits that the 1991 harvest has been good (and for any farmer in the world to admit such a thing, it means that his harvest must be exceptional). The markets are overflowing with grain, selling of course at very low prices that benefit .... the merchants. In the Sahel we follow a price index, which is known as the Goat-Millet Index. It is very simple: it tells you how many kilos of millet can be bought with one goat. Goats are savings. Next March, these very same markets will be crowded with sheep and goats and cattle for sale, some donkeys, maybe a few camels for sale. This will be March, in the middle of the dry season, when fodder is scarce and animals start to lose weight. The Hungry Season will be approaching; herders will be needing to sell a goat to buy some millet to feed their children. Millet prices will rise, livestock prices will fall: and a goat will buy less and less grain. This is when the merchants start to make a killing: they who bought cheap at 35Fcfa per kilo, will be able to sell at 120Fcfa or 130Fcfa, or perhaps even more. Last June in the village of Sarafere (eight hours from Timbuktu by canoe), the price reached 250 F per kilo, and the Village Elders met in Crisis Committee. At their behest, the Town Crier was sent around with his drum, to beg the merchants not to take their last golden earrings from the Fulani women: "We ask you, in the name of Allah and charity and fellowship, to accord them more credit, and not to accept their gold ornaments in exchange for millet. Hear ye! Hear ye!"

Did the merchants hear him? Well, they resorted (as a compromise to please Allah) to the time-honoured swindle of the pawnbroker. The Fulani women's gold earrings are in pawn, with usurious interest rates attached to them: the women are free to redeem it, but only when they can pay the compound interest as well as the original loan. Their last gold earrings are gone: a symbol of their former wealth yes, but also a symbol of their marriage, and the symbol of their status as free women in this land of slaves. The Fulani are numerous in this region south of Timbuktu, but they are not the only people who suffer from hunger. All those whose granaries are empty, are forced to beg for food. The young men go off to the Gulf of Guinea coastal cities (or to Libya) to become wage-slaves for survival. Those who remain have to find seeds to plant, when everything else has been eaten. Where do they go for seeds? What seed quality do they buy? They take whatever rubbish the merchants give them. What else can they do?

One solution to this dependence on money lenders is the creation of community-owned seed banks. I have been promoting such village-based cereal reserves for decades.
"Now tell me this," I asked a village meeting beside the river beneath a giant acacia tree: "When it comes to cereal prices, is the merchant stronger, or is the farmer stronger?" The old men gave out a bitter laugh, while the younger men behind chuckled at the foreigner's stupid question. "The merchant is always stronger" commented the President of the Village Association acidly.

This is the answer I was looking for. I continue: "Confronted by the merchants, a single farmer is as weak as a piece of straw. Is it not so?" The old men grunt in agreement. "If the President takes this piece of millet straw, he can break it easily, even though the President no longer has the strength of a lion as he used to have many years ago." Several of the Elders mutter their approval. The President nods, pleased both with this recognition of his age and seniority, and with the recollection of his youth. "If he takes ten pieces of straw together, he can still bend them, maybe even break them. But if he takes twenty, or thirty, or fifty pieces of straw... then who is stronger, the President or the straw?"

"We understand what you are saying" nods the President. "Together we are stronger. CARE has already told us this. We must create our own store of seeds, now at last that we have a good harvest."

CARE and its partners among the voluntary agencies are helping villagers to organise themselves. They must create and manage their own cereal banks, selecting the best grain for seed, and keeping the rest to feed the community through the hungry season. A cereal bank buys from the villagers, and then resells or lends to them at a profit. There needs to be a profit, in order to cover losses through waste and decay: but the villagers themselves decide on the profit. A profit measured in grain, and not profit that is turned into cash. A just profit: not usury. This is democracy in action: economic democracy.

"We must demand two sacks for every one sack loaned," cried a farmer in one village assembly I remember, where the agency ACORD was working with farmers along the bend of the Niger. "In this way our seed store will grow stronger, we will have more and more seeds in reserve each year; and we must check the grain in order to select the best seeds for planting the following year."

In addition to this seed selection, ACORD seeks out drought-resistant seeds from other areas of the Sahel, and brings in samples for other villagers to plant and try out. The drought years started back in 1965, and they suggest that the Sahel's climate is changing, becoming dryer and harsher. Finding new drought-resistant seeds may be the secret for survival in North Mali.

Seventy men in turbans were gathered in the dusty shade, and the farmers were debating how much interest to charge themselves. Two sacks for one means 100% interest, to be repaid in 3 months and in grain (after the harvest). In money terms that is negative interest: they will be borrowing one sack when prices are at 600Fcfa per sarwal in June, and repaying two when prices are down to 200Fcfa. If they converted this to cash as the merchants do, they would pay back three sarwal without counting interest: probably for every sack borrowed, they would pay four sacks = 200%, or twice as much as their cooperative seed bank. After the harvest when there is plenty of grain, farmers can afford to repay two sacks. 100% interest in grain over 3 months is less than half what the merchants charge. The villagers agreed enthusiastically to reimburse two bags of grain into their own cooperative seed bank.

HOPELESS LOVE

Our daughter Leïla is ruthless in her affairs. She ditched Eric because she thinks there is a 15-year-old who fancies her. Eric is a beautiful bronze color. His mother is from Chad and this makes a fine mix with his French dad. I gather that her new flame is about the same shade of bronze, although not a lot else is known about him. At least, not by me. Skin color is a vital part of mutual attraction in Africa. Light men and women are more easily attracted to each other while black men usually prefer black women; but some men go for a lighter skin than their own, which explains the market for those horrible Nigerian skin lightener creams that damage the epidermis. When I saw our other daughter Nana (elder daughter of our gardener Nafo and our clothes washer Awa) holding a tube of Nigerian skin lightener, I led her to the long-drop toilet and slowly squeezed the entire contents of the tube into the "negen" saying "Nana, you are a beautiful girl with a lovely skin, and never - in my house - will I allow you to destroy your beauty with these terrible creams."

How does Leïla know that the 15-year-old fancies her? Well, it happened while her class were playing handball (a game in which Leïla is an effective goalie, or so she says). When some seniors crossed the playground on the way to a physics lesson, this unknown admirer stopped to give Leïla "la bise."

"La bise" is a French air-peck on both cheeks. This moment of public intimacy took place shortly before Leïla's 12th birthday. To work out the chances for Eric being reinstated, I thought I had better study the nature and significance of "la bise." The charming French tradition allows symbolic kissing ladies on the cheek once, twice, thrice or even four times depending on which part of France you pretend to come from (or depending on how hungry you are feeling). You really get very little flesh contact from this air-peck: the chief benefit is a whiff of French perfume. As a tip for those who want to try getting away with a maximum of cheek-eating, Parisians only peck twice, whereas Bretons take two bites out of each cheek (or the air around it). As an avid "biseur" of pretty French women wearing fabulous perfume, I have adopted the meat-eating Breton tradition, nuzzling pretty feminine cheeks whenever I am allowed to while inhaling deeply. I love French perfume. Does this make me a "sniffeur" as well as a "biseur"?

Down in front of the French international school, I have not seen any sniffing, but there is plenty of bising as students arrive at 7.15 a.m. At school "la bise" takes place mainly between girls.... but only girls who are "friends." Others shake hands or do nothing. It seems to be only among the seniors that boys and girls give "la bise" on a regular basis.... again, it is between friends or family members. So, when Leïla received "la bise" from a fifteen-year-old, it certainly looks as if she is counted as a friend. But one peck on the cheek doesn't necessarily lead to pecks elsewhere (especially for nearly-12-year-olds!). Unless, of course, this is a 15-year-old predator.

I think (perhaps I hope) that Leïla may have read too much into her "bise." Chet says he knows the biseur, who said "hallo" to Leïla only because she is Chet's sister. That doesn't please Leïla, but it makes sense to Jeanne and me: what 15-year-old Rap-Fan is going to fancy a girl of eleven? Even if Leïla is a rather skilful Rap dancer: in West Africa, all girls know how to dance. Before they became interested in boys, Leïla used to spend whole weekends just dancing with Salima and her cousins under the mango trees. There is nothing to prove that the 15-year-old has even seen Leïla dance. But no one would expect Leïla to listen to parents! She believes she is in with a chance here and she decided to clear the decks of thirteen-year-old Eric to make space for his elder and better.

Leïla is not an especially arrogant girl. This admiration from the other side of the eros barrier is taken as perfectly natural. "I was elected Class Delegate, she explained, because all the boys voted for me as well as my friends. Only the French girls didn't vote for me, but I don't care because I hate them." (Leïla's mother Jeanne is French, but never mind.) Her election victory was based on swinging the male vote. Friends are naturally girls: no one can consider a mere boy as a "friend." Boys are either boyfriends, or they are consigned to the trash-heap of indifference. Leïla continued: "When the Teacher said that I had 13 votes out of 22 pupils, I heard a gasp.... and then I looked around and realised that all the boys are in love with me." We were eating fish-and-rice at supper when Leïla made this announcement. Jeanne giggled, I swallowed, and Chet sneered. Leïla was not at all put out: "It is true, and if you don't believe me, that is your problem." She helped herself to another piece of fish and the debate was closed.

I cannot tell whether my daughter really haunts the dreams of boys, but she found that she was not immune to sentimental regrets herself. Eric has recently been reinstated as a boyfriend, keeping the other at hand as an admirer. And there is a new candidate who rang up the other day, with a Canadian accent.... Chet must be quite impressed with his sister's influence at school for he announced at a family dinner party with friends, "my sister is now polyandrous." Where did he learn that word?

Leïla certainly seems to have an effect on one boy. I found a letter lying around from one of her twelve-year-old Malian classmates, which I have tried to translate for you. It starts with a heart drawn in red ink with LEÏLA written inside it. How touching. Well, here goes:

"My very dear Leïla,
Since my arrival in this school [they were in the same primary class together five years ago, so this is no sudden event], I have fallen hopelessly in love with your charming person. Since class 2 until now, I write you letters to proclaim my love for you. I have done everything to forget you [difficult, since he sits behind her in class every day] but my heart cannot resist either your beauty or your grace. I would have liked to make this declaration in tête-à-tête but I have never had an opportunity.

You know that a girl without a boy is like a rose without petals. It is for all these reasons that I wish to ask you if you will have the courage to go out with me. I await your response in the hope that you will accept. From your admirer XXX."

XXX is 12. Even in the original French, he sounds pompous! Should I nickname him "petal"? This lad is shy and plump and dark brown. I fear that he has no chance of winning peaches-and-cream Leïla in competition with bronze Eric and the 15-year-olds. "I have never had an opportunity," he writes. What a load of rubbish! They have been in the same class for five years. XXX even sees Leïla at weekends, since Salima is his cousin and she and Leïla virtually live as sisters. If XXX hasn't made it yet, he never will. It is a tough life, approaching adolescence and already cast upon the trash-heap of indifference.

HAPPY HUNGRY POLICE, A LETTER BY JEANNE LACVILLE

Everyone has his or her own policing tactic. Lacville pays. Now he will tell you that he is just being African. He'll tell you, "The policeman has to feed his children, so why not help him?" But the truth is: Lacville is a coward. Let me illustrate the Lacville easy-pay technique. We were driving around the Patrice Lumumba Square last month when a blue-uniformed policeman raised his hand. And his whistle. Lacville slowed to allow three laden bicycles to pass, avoided a donkey-cart full of leaves (sheep fodder ready for the next annual ovine slaughter) and pulled up beside the curb. The policeman sauntered up with a grin, and half saluted. "Les papiers?"

"You know perfectly well the papers are in order, you scoundrel," Lacville grinned amiably. "Do you really want to see my papers?" The policeman laughed and clapped his hands together in easy complicity. "Of course, I don't want to see your stupid papers. Oh, hallo, Madame, comment ça va? Et les enfants?" I said "Hallo," without enthusiasm. Last time he stopped me, this policeman asked me if he could marry my thirteen-year-old daughter (I think it was a joke, but Leïla was furious!) The policeman turned back to Lacville: "I don't want to see your papers: I just need 1000 francs." Lacville laughed cheerfully. "Well, the Eid fete was costly so I suppose I ought to help an old friend like you. Et la famille: Ça va?" Lacville reached into his pocket and pulled out some notes. The smallest was 2500 Fcfa. Lacville beckoned the police sergeant: "Eh camarade! Today is your good luck. But 2500 francs is too much for this rascal: you must share it." The sergeant took the money hastily: "No need to wave it around for the whole of Bamako to see." Lacville chuckled delightedly, and drove off with a wave.

A week later, the same policeman hailed me as I was driving home in the evening. I was tired, in a hurry to get back to the children. I was fed up with the office, annoyed with the phone bill, irritated by the heat. I pretended not to see the policeman's wave. "That is Lacville's policeman," I reasoned, "let him stew in the afternoon sun!" The policeman whistled. I pretended not to hear his whistle. In my rear view mirror, I saw him climb onto a moped, and set off in pursuit. Oh God! Quite unreasonably, I accelerated towards the bridge, swerving past a pickup overloaded with children. The rush-hour traffic can be quite a snarl on the Bridge of the Martyrs, and I hoped that the policeman's path would be as overloaded as the pickup. The cars moved slowly over the Niger River bridge. In the mirror I watched my policeman zigzagging after me.

"What am I going to do?" I wondered. By now I was clearly in breach of the Law. If it cost Lacville 2500 Fcfa just to say "hallo," what might be the fine for 1) refusing to stop, 2) not responding to a whistle, 3) fleeing from the police, 4) evading arrest.... I decided that I could NOT face it. Not tonight. I had an inspiration: if I could just get away tonight, Lacville could handle it in the morning! I would shelter behind my wifely status and leave it to the head of the family. I felt no feminist shame. Let Lacville easy-pay his way out of this one! Serve him right! I tried not to think what Lacville would say: just let them arrange things in a manly way "entre hommes."

At the exit to the bridge, I cut left, forcing a bus to brake sharply and causing an old man to fall off his bicycle. Ignoring the bumps, I hurled the car towards the second corner. As I turned right, I glanced over my shoulder. In my rush, I nearly crushed a game of football. And I saw the policeman turn left off the bridge in my direction.

Two corners later, I drew up behind a tree and hooted: would Carol's watchman open the gate before the policeman appeared? Too late: There he was at the end of my mirror.... I accelerated away towards the river. If I took the track round behind the United Nations office, through the riverside lettuce gardens, I could rejoin the main road and outpace Lacville's policeman along the tarmac. Oof! With a total disregard for my shock absorbers, I hurled the car towards the lettuce beds, and screeched around the bend in a swirl of dust. "I must slow down, or I'll crash the stupid car," I thought. I gained control of myself, and drove sensibly towards the tarmac road. Just as I reached it, Lacville's policeman rode up on his moped. I groaned. I rolled down the window, ready to surrender, feeling sick.

"Bonsoir Madame!" The policeman looked hot as he saluted. "You didn't see me at the Lumumba Square. I waved to you, and then I whistled, but you didn't hear me." I tried to look innocent and surprised, even as I composed my nerves for the next accusation. "So, I followed you. I have been following you ever since my post beside the Development Bank on the other side of the river."

Would Lacville ever forgive me? Mentally I was preparing myself for a night in jail. How would I ever live down the humiliation? The policeman wiped his brow, and smiled. "You see, I have news for you and your husband. My wife has given birth last night. She has given me twins."

A RESPECTABLE MARRIED WOMAN

Sheep and cattle need water. It is June and I am visiting Douentza, a dusty town beside the tarmac road in North Mali. Douentza has a big cattle market, where I enjoy mixing with the herders, evaluating livestock, and listening to the negotiations. In this hot, dry season, animals need lots of water. The Fulani nomads bring their herds to the river, or congregate close to ponds and lakes still holding water from the rainy season, which ended in September, nine hot months ago. Over Douentza's stagnant pool hangs a pall of dust. Flies and mosquitoes buzz around a thousand sheep and goats and cattle and donkeys, waiting restlessly for a buyer. A solitary camel is on offer: he chews contemptuously in the shade of a prickly acacia tree.

I push my way through a group of Tuaregs selling knives and bracelets, avoid the small black slave women with their plaited mats (because I already have a dozen such mats at home), and head for the smoke of grilling meat. The goat ribs look fresh enough, but the butcher is asking too much. A young girl with an enamel bowl on her head shows me her fried spicy chickens at two dollars each. She beckons me inside the nearby house, where her mother runs a restaurant. It is cool and quiet in here, under a straw shelter and out of the dust and heat.

Mrs. Cissé grills chickens, but she doesn't seem to have many customers. Her three daughters sell chicken pieces around the cattle market while she and her sister's daughter cook rice. The latter has a baby. She is not married, but no one cares about that in West Africa. Children are part of the family. They are welcomed and loved for themselves, and not for the sexual habits of their natural parents. Sexual rules exist here of course, but they are different from those in Europe and they also vary from place to place (and from family to family).

The chicken was tasty, so I said I'd come back in the evening for some of Mrs. Cissé's rice and stew. As the sun was setting, Mrs.

Cissé spread a mat for me and I stretched out to relax under God's desert sky. Every star leaps out at you, the tiny Pleiades are clear as buttons. The only other light is coming from the flickering fire beneath three iron pots. One daughter is pounding her tasteless millet paste, called toh. She lifts a branch from the flames to study her pot, and the sudden blaze lights up the small yard. Two cows and four sheep huddle closer together against the mud-brick wall.

Mrs. Cissé comes to sit on my mat. She is a handsome woman, who moves with the dignity of a nomad. The flame lights up her bare left shoulder. Her milk-chocolate skin is beautified with a large blue tatoo covering the whole of her lower lip. While I eat rice and sauce, she admires the gold embroidery around the ankles of my white African trousers. She tells me that she is the third wife of a café proprietor on the main road. It is not good to be an unmarried widow. A respectable lady should be married. If your late husband's brother doesn't marry you, you must find another husband to give you respectability.

The arrangement is that the two families keep separate. Once each week, Mrs. Cissé moves over to her husband's café for two nights, taking with her food for them both. That is her only obligation. He never comes to see visit her, and she feeds her own family. By now, Mrs. Cissé is admiring the gold embroidery on my cuffs and neckband. Unfortunately, as she explains, her daughters only go to sleep at 11 p.m. After that she is free. Having admired my beard, Mrs. Cissé starts massaging my feet and ankles. Massage in West Africa is a way of showing respect, or hospitality, or sometimes affection. Children massage their parents. Wives massage the feet of their husbands, or the feet of their husband's friends. Foot massage with cool water is especially welcome, because the feet become tired and dry with the dust of the Sahel. Since I bruised my toe on a rock earlier, her admiration of my feet is a little painful.

It is 10.15 p.m., and Mrs. Cissé is now admiring the shape of my left kneecap. The rice was very good, thank you. Delicious cooking. But I think it is time to pay my dues, and go home before Mrs. Cissé's hands go any higher.

RIVER BLINDNESS

In Europe, a blind person is symbolised by dark glasses and a white stick. Africans don't use white sticks. The symbol of blind people in West Africa is a horizontal stick held between the blind adult on one end and the little child leading him (or her). Child labour is normal in Africa. Sometimes I feel my kids ought to do more. Their life is so easy, so lazy except for school (which is pretty hard going, admittedly, when the temperature is in the high forties Celsius, pushing 120 degrees Fahrenheit).

In the village, daughters look after the young children, while little boys herd the goats or chase away birds and monkeys from the crops during the growing season. In the town, children push hand-carts, collect trash, sell mangoes and peanuts, or lead blind beggars. There are four blind people to whose survival I contribute about five times per week and another six or so to whom I make contributions when I meet them outside the bank, the petrol station, or the office. All of them are led by a son, a daughter, or a nephew aged around seven years. And most of them are blind from onchocerciasis.

River blindness is so-called because the oncho fly likes waterfalls. Actually, it is a sort of gnat (Simulium spp.) which transmits a miniscule worm. This worm wanders around the body, laying nests of eggs that form nodules. Eventually the worm (Onchocerca volvulus for entomologists and Latin scholars) may get around to laying eggs behind the eyes. This destroys the optic nerve and causes irreversible blindness. They say there are 20 million victims of this hideous parasite around the world, and they have my sympathy.

The best person to tell you about a disease is not the doctor, but the victim. A friend of mine had oncho some years ago and she says it is hell. Never did she believe that itching could lead to suicide: but the constant itching in her arms, shoulders, elbows, wrists, legs, calves, and ankles turned her into a helpless gibbering wreck with limbs bleeding from scratching. She was ready to try anything, even suicide. In the end the doctors treated her with arsenic. Arsenic is a poison, but she took very, very small doses: she went from a single tiny pill of 0.26 grams and built up to 125 grams per day. It took a year. It killed the filarial worms. It also left her with such a high seral arsenic build-up that she is forbidden ever to go near any bloodbank for fear that she might kill the next person with arsenic poisoning.

There is no vaccine against river blindness. The only prevention, it to avoid being bitten by the fly (or gnat).

Professor Warren Berggren of Harvard tells a tragic story from Burkina (formerly Upper Volta) about the protection of villagers from river blindness. There the World Health Organization (WHO) has been campaigning for decades to eliminate river blindness. In this case they moved villagers away from the flowing river areas and settled them on drier lands where the oncho fly never comes. But the young men went back to work their fields where the risk of infection was so very high. Why did they take such a dreadful risk? They explained: "We must earn money very quickly so that we may marry young. This is necessary in our culture because we need to have babies quickly. It is the only way to survive later, in ten years' time, for we must be sure of having young children to lead us then, when we are blind."

Yet the WHO program has been pretty successful. Admittedly at huge cost, the light planes have been spraying rivers across West Africa to kill the Simulii and helicopters have been zigzagging around the cliffs, hovering over waterfalls to spray the breeding havens of these nasty little flies. Oncho has been going down. But they never get all the flies. For twenty years they were never able to fly into Guinea at all as long as the President Sekou Toure ran the country into ruins. While many areas of the Volta River region have been cleared, there are a million other foyers of oncho still making people blind.

But here comes medical science with another breakthrough. In the past four years or so, they have discovered a drug that kills the Simulii in the human body. It is an American antiparasitic drug sold as Mectezan, or Ivermectin, which has been on the market for 27 years for treating heartworm in dogs. Ivermectin originated from a single microorganism isolated in 1974 at the Kitasato Intitute in Japan from local soil. Later, someone discovered that it works against the river blindness worm, not by killing it, but by reducing the release of larvae from the adult worm.

Better still: Merck, Sharp & Dohme, the company that sells Ivermectin, has been making so many millions of dollars from it for so many years (with annual sales in excess of over US$1 billion), that they are now offering the drug free to treat river blindness.
Yes, it's free!

Now I know about this; and thanks to this article, you know about it too. But the people who need the drug do not know about it, and they are never likely to read the Guardian Weekly (despite its huge circulation across the world). So how in heaven can we get this wonderful drug to the remotest villages where the oncho fly is working away so assiduously, laying its eggs? The World Bank? Forget it! 90% of World Bank staff live in Washington and they never get closer to the villagers than a five-star hotel in the capital city. Governments? Most governments in Africa have small budgets and their medical services are centered on regional or local capitals. If they had a mobile oncho team, they might be able to distribute the medicines, but they don't, and they won't.

No, this one is a job for the voluntary organisations: the charities, the missions, the non-governmental organisations. Most of their work is carried out in distant rural areas. They have the desire and the means to reach the remotest villagers. Now the question is, how can we get the free drug to the right voluntary organisations, to the people who know how to reach remote villagers?

To that question, I admit I have not found an answer.
But other people are searching for answers.

Note in 2024 on progress: In 1995, WHO launched the African Programme for Onchocerciasis Control (APOC), using Ivermectin to eliminate onchocerciasis from African countries in which the disease was endemic. No African country had eliminated the disease when APOC ended in 2015, to be replaced by the WHO Expanded Special Programme for the Elimination of Neglected Tropical Diseases (ESPEN), which works with health ministries and non-governmental organisations to achieve treatment of 80% of the population in endemic areas for at least 10–12 years, the life span of the adult worm. According to the World Health Organization, the four countries that have eradicated onchocerciasis are Colombia (2013), Ecuador (2014), Mexico (2015), and Guatemala (2016), thanks to mass administration of Ivermectin.

SUGAR AND FRESH LEMON JUICE

Salima came around to see me yesterday, on an urgent mission. Salima is the daughter of my "Old Brother," my Koroké, and she is also best friend to my daughter Leïla. Well, actually she is not best friend this week since they are both eleven. Leïla had a SECRET and Salima betrayed the secret to Fatou and Marie in the playground, so she is not on speaking terms with Salima this afternoon. In fact, she is actually sitting at my typewriter at the moment, laboriously thumping out a letter to Salima full of threats about repaying deceit with treachery by revealing some of Salima's SECRETS to some other little horror named Magalie. But normally they are very close. I expect they'll be playing Cindy dolls again this evening. In African terms they are "sisters": they play together, skip together, dance together, eat and sleep and whisper together, and go to parties together but only if Philippe isn't there (presumably another of their class enemies).

Anyway, Salima's mission was more serious than all this. She said she had to speak to me. I bent my head to hear her, and got a kiss on the cheek for my pains. This put me in a very good mood. What could I do for Salima? Well, Salima had been over at her auntie's, and one of the cooks there has a little daughter called Oumou who had severe conjunctivitis, and so her mother had squeezed lemon juice into her eyes and then added some granulated sugar. And little Oumou was crying, and Salima wanted me to look at her eyes.

Why in the name of Allah had the mother put lemon juice and sugar into her daughter's eyes? Well, the child was crying, and she had no medicine, and so she used what she had. The mother would have done better to pee in her daughter's face! I'm not joking either. Urine is usually pure, and warm, and mildly antiseptic (the weak uric acid is helpful). In the desert they use urine for wounds and infections. It soothes, it washes the gritty sand out, and it is antiseptic. A man I know was surprised, when he cut his foot in southern Algeria, to be told to put his foot into an empty bucket. He was even more surprised when his three Tuareg companions proceeded to pull out their you-know-whats and fill the bucket with their combined urine. His third surprise was to discover that this homely remedy worked well: his foot healed up quickly and cleanly.

Still, there are alternatives. I am not necessarily pushing readers to try desperate desert remedies. We usually recommend warm water with a little salt, or cold tea: in both cases the water is boiled, then allowed to cool. Salt is a better disinfectant, but tea is safer because you can be sure the water was boiled. The tannin in tea seems to be soothing for the eyes. Cold green tea is said to be best. A warm tea bag also makes a good compress for a tired or swollen eye (contact lens wearers should make a note of this helpful information). Tea is my standard eyewash.

Little Oumou turned out to be about three. She was sitting silent in a corner on a small wooden stool, her wrists held over her eyes. Sand was encrusted on her legs, her back was brown with dirt, and when I looked at her, her face was a soupy mixture of tears and dust and pus. In Africa, as in Europe, street kids like Oumou don't get much cleanliness. There was no sign of the mother, so I told Salima to pick her up. That broke the silence! Oumou wailed and screamed and rolled in the earth. Two women stirring a vast three-legged pot over the fire looked on indifferently. So, I picked up little Oumou myself and carried her screaming and writhing the two hundred yards to our bathroom. There I sat on the edge of the bath with Oumou on my knees and calmed her while Jeanne went off to find the eye ointment. Salima stroked Oumou's hand, and I stroked the back of her head, holding her tightly against my chest. By the time Jeanne returned from the medicine cupboard, the little girl was rested and quiet.

After that it was straightforward. I changed my shirt. Jeanne and Salima washed Oumou, sending husband (that's me) off to boil the traditional kettle. Then I mixed the warm water salt solution while Jeanne gently wiped the infant's eyes with clean tissues. Ten minutes after arrival, Oumou was already looking around with one clear eye and we were able to see how the other was red and inflamed. But there was nothing very wrong with her eyes. Tens of thousands of people have dust-irritated eyes in the Sahel. Not all of them get the sugar-lemon treatment. Most of them get no treatment at all.

It is ironic, for the government here has made huge progress in health and medicines. Despite the deliberate and persistent sabotage of the French and Swiss pharmaceutical firms, a national policy has been established to provide cheap basic medicines. Mali is one small country fighting to achieve medical progress for all. Instead of imported branded products in expensive packaging, generic products are being made here at a tenth the price. While the French Government is obstructive, protecting its pharmaceutical corporations, the Chinese government has financed and built a pharmaceutical factory in the capital city. It's not all roses, since chemical effluent is now polluting the Niger River, but it is medical progress. Among other products on general sale, there is an aureomycine antibiotic eye ointment in a tube, which costs 130 Fcfa = 22 pence (half a dollar). Even a rural maid can probably afford that for her child's eyes.

So why didn't Oumou's mother? That's easy: the lady doesn't know about the eye ointment. There are almost no public campaigns for cleanliness. We live on Old Brother's fruit farm with a whole network of workers and maids and inherited household slaves and hangers-on, none of whom knows about disinfectant, or the importance of killing germs. Old Brother's family all know of course: his sisters and brothers and cousins are highly educated, many of them ambassadors or lawyers or senior military officers. Little Oumou's close family contains no one who has ever been to school, therefore they have not benefitted from hygiene lessons. The national average for primary school attendance in Mali is 23%.

Few Malians learn about cleanliness. A village housewife in the remotest village will sweep her yard meticulously every morning, removing every leaf and twig and scrap of animal dung; but she knows nothing about microbes, bacteria or viruses. Everybody listens to the radio, but they hear mainly politics and music. Very little education is provided on the radio. Knowledge is the first requirement for good health. The doctors are perfectly competent in Africa (although there are not enough of them). Plenty of people here get cured of their sickness. The problem is that so many of them get sick unnecessarily.

Surprisingly enough, the drought years have helped. Thousands of tube wells have been sunk by the United Nations and other agencies. Tube wells bring up clean water through pipes and pumps. That doesn't mean the water is always clean, of course. Those that were paid for by the fanatical Saudi government, have been sunk by beer-swilling German technicians who never speak to a villager – they offered no hygiene training. Put clean water in a dirty bucket and you have infected water. Leave clean water all day in an open container, and it will become cloudy with dust and dirt. Dip a dirty cup into the family water jar, and you may infect the whole family water supply with your microbes.

In addition to the tube wells, there have been thousands of hand wells dug throughout the Sahel. Many more have been dug by villagers themselves, under local supervision. If these supervisors are from NGOs, they are usually keen on hygiene training. Local government inspectors sometimes test and disinfect wells, explaining to the villagers the need to chlorinate their well every year. Local associations and non-governmental organisations run classes on hygiene for mothers and grandmothers. Well-meaning volunteers campaign for villagers to filter water, to disinfect clay water jars every week, to avoid throwing dirty buckets into the wells. More and more wells are equipped with walls to reduce erosion, with pulleys to facilitate levitation, with cement aprons to reduce dirt .... and with cement basins built twenty yards away from well in order to provide drinking troughs for the animals and wash-places for the weekly linen - far enough from the wellhead to avoid pollution.

In villages on the Sikasso frontier, I found that the NGO Save The Children had convinced villagers to surround their wells with clean gravel protected by a wooden fence that kept children and livestock away from the wellhead. Here was proof that Save has had success with hygiene training and public health messaging.

These days the dirtiest conditions may well be in the towns: sprawling areas of village housing sprout up with no urban planning and without the advantage of air and space and social organisation typical of life in a village. Many slums have poor wells, with private cess-pits polluting the local water table. That often leads to fecal contamination of drinking water. Add negligible incomes, poor food, widespread malnutrition, miserable schooling, divided families, and you have built an image of extreme urban poverty. The disruption of the extended family, without the proximity of grandparents, often means that modern ignorance is compounded by ignorance of the traditional rules that promoted health and hygiene. In many villages, the old people know herbs to cure sores, to heal wounds. But when their offspring are stranded in the city, they turn to modern remedies that never existed in the village. Like sugar and lemon juice. Ow!

DOGON ONION BALLS

Bandiagara is a dry town on a stony plateau. Here there are no peaks. These are ancient rocks in the border lands between Mali and Burkina Faso, rocks that have worn by the wind into weird spiritual shapes. There is not much soil on the Bandiagara Plateau. The scattered trees are leafless and lifeless until the rains come in June. The Dogons moved here centuries ago. To evade war and forced conversion to Islam, they came to live in a rocky landscape where they could not be followed and where they just manage to survive. Dogons scrape a living from among the stones.

Dogon farmers stand high in the league of the world's most careful cultivators. Every scrap of earth is cultivated. On this wind-swept plateau, farmers heap the thin soil around each millet stalk to help it find sustenance to grow. A Dogon "field" may be a hole between two rocks where there is space to plant just half a dozen millet stalks. In crevasses and along steep slopes, the Dogon farmers grow their food. Where the rainwater races down ancient gulleys, the Dogons have built stone barrages to catch the water in a pond, reducing erosion and holding the run-off. On the flat patches around the pond, they grow onions, which need more water than millet. Small Dogon onion plants are packed tightly together in neat rows. They are fed with pounded goat manure and water - lots of water in this hot dry atmosphere where the sun sucks moisture out of every living thing.

Through the heat haze mirage, onion fields glisten shining green like emeralds among the grey tortured rocks. Between the stones, farmers scrape together some earth and plant an onion or two. In front of one small house, I found a cairn of stones with a bright green patch of hair. Coming nearer, I discovered that this too was a field of onions. The owner had built the cairn and collected earth by hand to create another miniature bed to plant his precious crop of onions.

The onions are truly precious. In recent years, onions have provided the Dogon farmers with a cash crop. Previously their only local export crop was people. Young men left the villages to find work to feed their ageing parents. A job in the town became the only way for a young man to afford a wife, to raise a family. Often the young men had to travel to earn the cash for their parents to pay taxes: for many villagers, their only need for cash (even in the 1990s) is to pay taxes, and the fines that are imposed upon them every year by the vulturine administration.

Onions have developed trade. Trucks full of dried onions travel down from the Bandiagara Plateau to local and regional markets. The onion leaves are crushed together into green balls of flavouring for soups and stews. The onion balls are dried in the sun, and they keep forever. Sacks full of onion balls decorate every market in the Sahel, and the trade is largely dominated by the Dogon people themselves, who earn the profits and take them back to their villages.

Their major competitors are the cube manufacturers: purveyors of monosodium glutamate led by the ubiquitous Maggi cubes. These imported cubes are the opposite of the Dogon onion balls. Onion balls are locally produced, natural, cheap, and create indigenous wealth. Cubes are chemical, industrial, and expensive. Cubes earn profits, which are sent outside the country to pay for patents and marketing costs, expensive European machinery to mould them, and imported chemicals to flavour them.

Now a new threat has emerged to the onion ball economy: overproduction. While Maggi and all the other cubes whittle away at urban market share ("modern housewives prefer the flavour of delicious imported monosodium glutamate" sing the television ads), other regions have jumped on the onion bandwagon. Dogons are selling more onions, but prices are falling.

As so often happens in the primary economy, others can easily copy a good idea. Imitation is flattery, but it can easily kill the golden goose. That happened with coffee: Latin America has replaced Africa as the world's largest coffee-producing region and the African share of the market has fallen from a peak of 34% to barely 20%. It has happened with oils: American soya bean and Asian palm oil have reduced Africa's market share from 73% to 27%. And it is happening with onion balls as the people of the Niger River basin attack the Dogons in the markets of Mali, Ivory Coast, and Burkina Faso.

Guido Kap is a Dutch volunteer, helping villagers to build small dams with the Catholic Mission. "They are never going to make out with their onions," he says. "Prices are falling in the markets, and we are still building more dams for more Dogon villages who also want to grow onions. They cannot eat the onions: onions are a cash crop. With the cash value of onions falling, the Dogons must start using the dams to grow food, and find another source of money to pay their taxes."

And he points to the Asian example. "What we need now is some systems to make value from the water. Sure, when the Dogons are starving, they dig up the roots of the water lilies: but they have got to do better than that. They need to go in for fish-farming. They need carp and tilapia, which feed on weed and plankton to keep the food-chain short. In China they keep also ducks, which enrich the water. That way the ducks help grow more weed to feed more fish. These are ideas that have been used around the rice fields of Asia for centuries. How do we introduce these concepts to Dogon farmers who have never even had water before? How do we get them to adopt such ideas to help themselves?"

That is the fundamental question for hungry Africa. Techniques do exist for feeding the world's population. How do we bring the techniques into the farms of the world's hungry? The World Bank has proved that you cannot develop people simply by throwing money at them. In fact the World Bank has been largely responsible for sacrificing food crops in order to develop cash crops in Asia that have impoverished Africa by driving down prices, thereby increasing Western corporate profits in the private sector. Centralised state economies in the public sector have been found wanting. There must be another way.

The charitable agencies within "civil society" are showing that you can only develop people by offering people the opportunity to develop their community themselves. The other "third way" is different from the private economy and the public economy: we need to promote cooperative efforts that will expand the "social economy" and build wealth in the communities from the bottom up. That is what we mean by "grass-roots development." In Dogon country there is no grass. Up there on the dusty plateau of Bandiagara, their development processes have to start from the onion roots.

DIAMONDS, BLOOD, AND CEMENT IN SIERRA LEONE

Sierra Leone is one of the world's biggest producers of diamonds. People call them "blood diamonds" because of the civil war and the killing. And the corruption. Sierra Leone became so broke that even the banks ran out of money. Have you heard about the country where the unit of currency was the bottle of beer? That was Sierra Leone in the 1990s, when there was so little cash available that beer became the only acceptable medium of universal exchange. During the 1980s, it was said that the country with the highest per capita import rate of new Mercedes cars was ..... yes, you guessed it: Sierra Leone. [The country with the world record for per capita champagne imports at that time, was Gabon, which also has lots of poverty, a country that sells petroleum to and for the French.]

Sierra Leone was bankrupt; but there were plenty of rich people, and you'd better believe it. Most rich people are not found in Africa's manufacturing sector. Sierra Leone has virtually no manufacturing sector. Like most of West Africa, the money is in trade.

Traders do all right, largely through smuggling. Where are the diamonds? They are in Amsterdam. How did they get there? Well, very few diamonds go through channels that are registered and pay taxes.

"There are plenty of ways to pass diamonds" chuckled a friend of mine to whom I posed the question. "You must come with me some time, Rob, and see how we do it. It is a great laugh. Now think, how many customs men will search under the flowing robes of an older woman? They may want to feel around, and not only for seeking diamonds: but they don't dare to do it. And do you think the customs men will search inside coffins? Eee, they are too respectful of the spirits to risk that!

Have you ever thought how many diamonds you can hide inside a good big bowl of cooked rice? My boy, there are so many ways to smuggle these sweet diamonds!" He clapped me warmly on the shoulder, as his several chins rippled with laughter at the stupidity (and cupidity) of customs officers.

"And if they catch you, then you negotiate. They always want something for themselves. The customs boys cannot risk demanding too much, for if we complain they could get into a big fix. They assume we know their boss. So, we negotiate." My friend roared with laughter at the venality of customs officials …. and at his own cunning.

Diamonds have played a big part in the history of Sierra Leone. Back in 1960s, an army general overthrew the civilian government. Then the colonels overthrew the general. Soon the lieutenants overthrew the colonels. Finally the sergeants overthrew the lieutenants and invited the trade unionist politician Siaka Stevens back from exile in Guinea. Siaka Stevens had been Mayor of Freetown. He was hungry for power, and hungry for diamonds. History shows that he achieved both his ambitions.

How was it paid for? Always with Diamonds. The final coup d'état was financed by Lebanese merchants who advanced money against payment in diamonds. Shortly after Stevens gained power, the national diamond company was instructed to load up a large quantity of diamonds into a small plane in up-country Kenema, and to take off for Freetown, the capital. Once in the air, the men who had loaded the diamonds produced firearms and informed the pilot that his flight plan had been changed: he should fly north. A small airstrip was pointed out to the pilot, where he should land. Two Landrovers drove out from under the trees, loaded up the diamonds, and disappeared with the armed hijackers into the forest. The national mining company was informed about an unfortunate and unexpected armed robbery. Payment had been made.

The Sahel is full of diamond profits. In Senegal and Guinea and Liberia, in Gambia, Mali, Burkina and Ivory Coast, the richest men include a handful of diamond dealers. Most of them have spent time in Gabon or Zaire, or in Sierra Leone. They deal mainly on the ground with their own nationals: the merchants buy diamonds from adventurous young men who try their luck at digging, earning enough to return home with a motor pump, a shiny radio and some plastic armchairs. These digging pioneers sell their diamonds to their trading compatriots, and the traders get very rich by smuggling the diamonds to Amsterdam. Where the Dutch get even richer......

Traders come in every shape and form. Many of them wear suits and drive Mercedes. In Mali I know one grizzled old man called Camara who wears a simple cotton shift and a round bonnet, and who looks like anybody's night watchman. Camara owns dozens of buildings in several capital cities, and he seldom has less than half a million dollars in spare cash. Another corrupt old man called Cissoko decided to spend his diamond dollars in the time-honoured tradition of returning prestige to his village. But he has gone a bit over the top: instead of building a villa for himself and a concrete mosque for the village, Cissoko decided to feed and clothe every man woman and child in the village at his own expense. Unsurprisingly, they elected him Mayor of their community. Talk about "creating dependency": even the World Bank and the French Embassy have never achieved 100 per cent dependency like Mr. Cissoko. Villagers no longer grow millet in that village. All the children go to school, but none of them works. If they have not learned the value of work, the boys will probably become urban lay-abouts. Oh well, Mr. Cissoko will no longer be around to see his failure and to say incredulously (like many parents): "But I gave them everything...."

Then there is a man called Diawara who owns loss-making hotels. Why does he invest his money so liberally in loss-makers? The hotels are not supposed to be losing money; it is just that Diawara doesn't know anything about the hotel business and won't hire a manager who does. Diamond smugglers are very suspicious of delegating responsibility. Mr. Diawara keeps in his own pocket the keys of the bars for all his hotels. If the boss isn't present, the bar cannot open! Even making losses, his hotels are useful because he is able to employ large numbers of cousins and nephews. This hotel employment system brings him family prestige, like Cissoko's village enterprise. Rumour has it that the hotels are also a useful front for smuggling other, nastier commodities, even more lucrative than the diamonds that funded them initially.

You might think that all these rich people would be helping to develop their countries. Why has diamond-rich Sierra Leone never become a hub for manufacturing? After all, capital accumulation is necessary for investment, and investment is necessary for production. If therefore the diamond smugglers and other traders have capital accumulation (and they have it in great abundance), it should be a good thing. The trouble is, they do not invest in a productive economy.

Where there is investment, West African Big Men invest in building mosques or churches or huge houses. These increase the consumption of cement (imported) and decorations (neon lights, plastic tiles, glass windows, metal door frames: all imported). Electricity consumption rises (diesel generators are imported; the diesel fuel is imported). There is no creation of wealth. The returns on investment to the individual are high in prestige, and also in cash from the house rents paid by embassies and other international organisations: but they do not produce wealth for the country.

There is an economic concept called "The Multiplier Effect" that describes wealth creation: a man who invests (say) in a bicycle factory can make a profit from selling bicycles. The people who work for him spend money and some of them may create enterprises. The bicycle factory will need to buy things like bicycle parts, work clothes, food and soap for workers, all of which will create employment and encourage wealth creation.

Workers in the factory will develop new skills. More bicycle owners will need more bicycle repair shops, and down-the-road the new skills may lead to wealth creation and new jobs like making bicycle-powered water pumps, creating tricycles for trading or carrying goods …. and increased trade also leads to increased wealth. This example shows how an investment in making bicycles can multiply its initial impact, and launch a new technical and economic ecosystem. It is one example of The Multiplier Effect.

Money for The Multiplier Effect" in Sierra Leone is squirreled away in British or Lebanese banks. The wealthy traders of West Africa like to keep their assets liquid. They spread wealth liberally in gifts and bribes, but they make no productive investments and they pay few taxes. No customs duties accrue from illegal trade in diamonds and gold…. often no even from their cement imports, which are whisked through duty-free with the compliance of hungry customs officers. In fact, if you study the list of house owners in West Africa's capital cities, you will find that, after the diamond traders, drug smugglers and ex-Ministers of Public Works, it is the colonels and customs officers who own the most houses.

PEANUT BUTTER GRINDER

"We are six poor widows. For me, my husband go long time to find work in Lagos. For three years I hear nothing. Then a man returning from Lagos to Timbuktu, he knew him. He told me my man died in a construction accident. We are all in poverty, and we are the only support for our families." Penda is a heavy woman, her forty-year-old black skin shining with sweat beneath her blue and orange headscarf. She speaks slowly, but with confidence. None of the older grey women will say a word. I wouldn't be able to understand them anyhow since I speak no Songhoy; but beside me on the mat is a charming bearded Tuareg, Ali Ag Abdou, who is kindly whispering into my ear a simultaneous translation.

Penda is still speaking steadily. "We have founded a group of six women to help ourselves. We are used to growing vegetables during the dry season. We dig the wells with our own hands, and we cultivate lettuces and tomatoes and eggplants and bitter tomatoes..... But now in the town everyone is growing vegetables. The prices are not good. We are not able to feed our children.

"Therefore we have decided to make peanut paste. Vegetables and salads do not keep well in this hot season. We cannot carry them to market. The river port of Tonka is twenty kilometers from our village. Vegetables cannot be good when we have carried them so far on our heads. But peanut paste is easy to preserve, it is easy to carry. It is easy to sell. Yes, if we can get a hand mill for peanut paste, then we will be able to earn money to feed our families."

Descendants of a great medieval Niger River empire, the Songhoy are rice farmers. But not the women! Unlike many African women, Songhoy women do little work outside the house. Although they rarely participate in meetings with the village men, Songhoy women have great influence in the family.

A Songhoy man dreads his wife's displeasure. If ever his wife should decide to leave him, he will become a laughing stock: for the bride brings with her to the marriage, the frame and the colourful woven mats with which she builds the home. If she walks out, she takes her mats with her! A divorcing Songhoy woman leaves the husband sitting foolishly in the sun, forced to crawl back to the shade of his mother's hut while his friends mock him.

Here we are, sitting under one such shelter built of reed mats attached to a frame of branches. One of the six women is sitting outside in the sun. Confronted by unknown men, she is too shy to sit with us. We have come in answer to a request for development assistance. This women's cooperative wants to set up a business. The nearest bank is 85 kilometers away in Timbuktu. How could they hope to visit Timbuktu? In any case, these women own nothing that could interest a banker, so they could never in a million years hope to get a bank loan.

Credit schemes for poor women do exist in the capital city Bamako, funded by the United Nations and by the EEC, but none of these women has even been as far as the regional capital of Timbuktu, far less undertaken the ten-hour journey to Bamako. Only non-governmental organisations like Save the Children, Oxfam, Christian Aid (or - as in this case - their local African partner NGO) ever reach out to people living in remote rural villages. If we do decide to recommend a loan or a grant to this women's group, they'll need to borrow less than 200 pounds sterling ($300 US), and no traditional banker is likely to be interested in that size of loan. They would see it as pitiful. Bankers look for financial returns; we are looking at social returns, economic returns, entrepreneurial returns, appropriate technological returns, humanitarian returns, and anti-starvation returns. These are what concern villagers. Financial returns only interest city folk who already have finance.

If these women are to succeed in their peanut butter enterprise, they have got to understand the project from the very start. Ali strokes his beard and stands up. He has been chosen as public scribe, both because he is a respected teacher and member of the Malian development association (a partner of Oxfam), and also because he is literate in Songhoy. We are all literate in the colonial language, but not many educated Africans can actually read and write their own language. That is part of the development process we are promoting.

Ali knocks the ash out of his slender Tuareg pipe and shuffles across the sand to the blackboard. He selects red chalk and carefully inscribes the date: we can tell that Ali has been a primary school teacher for thirty years. When Ali sees the old lady Fatoumata still sitting outside, he calls her to come in. Reluctantly she moves her stool inside the hut but keeps her back to the men. We take fifteen minutes discussing a project title: *tiga dege hinsa*, which means "producing peanut paste." The debate is in Songhoy. Fatoumata is interested and she turns around. What are the objectives of the project? How indeed can we translate the word "objectives"? Several people offer words while Fatoumata rocks slowly back and forth in concentration. Suddenly her high voice shrills out: "*Tenje diyo*."

Everybody looks at her and she hides her face in shame behind her faded green headscarf. Women are not supposed to shout out in a men's meeting. But here it is different. "That's right!" cries Ali in delight. He writes it carefully on the blackboard and Fatoumata shines with joy. She cannot read, yet she is actually joining in with literate people. Ali the Teacher has praised her.

From that moment, Fatoumata lost her shyness. Half the ideas on the blackboard came from her, as she became more and more excited about the project. Her project. From there onwards the meeting was plain sailing. We constructed the aims and objectives of the project. We worked out with the women the results they hoped to achieve. "We want to have several machines, explained Penda, but we must start with one machine first. We cannot manage a diesel mill for maize and millet until we have experience with the hand mill for peanuts. But if we are successful, we will earn money to repair the peanut mill and save enough to ask you for a new credit for a grain mill."

Then we all worked out the budget. We were able to approve the credit for their mill. They should be starting next week.

WHO OWNS OUR VILLAGE AFTER INDEPENDENCE?

A rich city friend has invited me out to his new field in the countryside for Sunday lunch. Let's call him Simaga. I agreed to meet him at his massive town house in Bamako, Mali's capital city. His cement-block villa is an extravagant mix of German, French and Italian styles, topped with red tiles and decorated with cement swans. We climb into his 4-wheel-drive vehicle and his driver takes us out of town.

After forty miles, we turn off the tarmac, and bump for a while across the ruts of a farm track. We pass a village where small boys are keeping a flock of goats away from the millet fields, and turn into a mango grove. A picnic has been prepared for our arrival. Simaga takes off his outer robe, heavily embroidered with gold thread, and hangs it carefully on a branch. We throw ourselves down on mattresses to relax in the shade of a spreading mango tree.

Simaga is justly proud of his country estate. Mango trees in Mali have neatly trimmed leaf canopies, thanks to the goats that stand on their hind legs to eat the low-hanging leaves. The trees are rich and ripe, the shade is deep and cool. "It is only this year that I found the site I wanted," Simaga tells me, as we suck fat juicy mangoes. "I was looking for an orchard of mature mango trees and cashew trees, beside a rainy season pond where I can get decent crops and grow some vegetables."

I congratulated him, and asked why the village had sold such valuable land. "Oh, they didn't sell it: don't you know that all land belongs to the State? The new legislation allows anybody to register land. The villagers don't use this land properly. They don't even grow vegetables here, although the city market is available. I have registered the land in my name, and I shall make it pay properly. But I like to be a good neighbour. I have already delivered fifteen sacks of millet, which I presented to the Head of the village in a public meeting." Simaga slurps another lump of mango into his chubby mouth. "They were all very happy with me."

Lunch started with charcoal-grilled lamb. "I have killed a sheep in your honour," said my friend. The lamb was followed by charcoal-grilled chicken; then we were served fish with vegetables and rice. The meal ended with a profusion of fruits: mangoes from Mali, bananas and pineapple from Ivory Coast, apples from France, and sweet Solo papayas grown here from imported seed. But somehow it didn't taste so good, as I pondered the plundered gardens in which I was sitting. How can a wealthy city magnate come and "steal" the best lands from these villagers, simply by registering their gardens in his name? I thought village lands were inalienable. There must be more to this than meets the eye. I decided I had better talk to Konaté, a Malian anthropologist friend who has studied matters of land.

"Of course, you are right," Konaté told me. "In the African conception of the universe, a man doesn't own land. It is the land that owns men. The first arrival is the possessor of the land, but not the owner: he receives the land from God The Creator. The man who cuts the bush to grow the first crops, he becomes the Chief of the Soil .... he and his descendants. He will allocate fields to other farmers. He cannot refuse to allocate land to newcomers, as long as there is land available: for in African culture, a man who is refused land is dead. What the Chief of the Soil really does is "lend" the land. A field that is no longer needed can be reallocated to another person, with the approval of the Chief of the Soil. But it is quite impossible for the land to leave the village."

Konaté's house is quite unlike the cement fortress built by Simaga. Konaté has spent ten years saving for his two-bedroom house on a new estate in a Bamako city suburb sponsored by the World Bank. It took him three years to get off the waiting list, onto the allocation list. "Finally, after I had spent more than a thousand dollars in bribes, an uncle of mine was promoted to Minister, and he succeeded in getting my application approved. That is one of the problems in Africa. In the village, traditional values survive. But in the city, we have adopted the worst habits of our former colonial masters, without any of the safeguards. Do you think a judge here is important? Any sergeant in the army can make a Judge jump like his string puppet. The only people who got houses allocated on this World Bank project without paying big bribes, were officers and non-commissioned officers in the army and the police."

Unlike Africans, Europeans and Americans are obsessed with ownership. An African is born into a family, the family is part of a village, and the village is part of a communal cultural inheritance. It is the family and the village that give identity to a person in Africa. In Europe it is the individual who counts, and his possessions give him status. Roman Law recognizes absolute sovereignty over land. In Roman Law (and in Napoleon Bonaparte's law), everything is registered: the French arrived in West Africa with their guns and their identity cards, and they imposed their Napoleonic laws. From the beginning of the 20th century until around 1934, they took whatever lands pleased them.

In 1935 the French government began to limit the greed of the white colonisers. Grand Customary Rights were recognised. The Chief of the Soil once again gained control over the land spaces cultivated by his ancestors. Of course, some lands had actually been conquered by military force: the nineteenth century in West Africa was a period of war and rape and anarchy. And slavery. Anyone who bought guns from the French or the British, immediately became more powerful than his neighbours. One of the first to gain such advantage was Sekou Omar Tall, a fanatical Islamic crusader whose cavalry swept down from the Senegalese mountains and sowed hatred wherever they rode. Tall was a jihadist using French firearms. His descendants revere him as the saint El Haj Oumar Tall.

In West Africa, the best protection against guns was magic. Many a medicine man got rich by selling juju charm protections against death from guns. One great advantage of this business model, is that disappointed customers are unable to come and complain. The anarchy created by 19th century Islamists and fomented by European firearms merchants was so harsh, that many Africans were relieved when the French took over during the 1880s: the French soldiers were brutal in conquest, but White rule at least brought some peace and stability.

Independence came in 1960. New African States adopted various anti-colonial ideologies: some stayed as de facto French or British satellites, while others moved under Russia's influence, or adopted some form of African Socialism. The new governments all nationalized the land in their new Nation State. First the people had owned the land, and the Chief of the Soil allocated parcels of land. Now the State was the land owner, meaning that land distribution was controlled by venal civil servants who responded to the best payers.

Africa's agricultural ecology for the past thousands of years was based on long fallow periods between crops: thin soils need time to recover their fertility. Under Mali's new legislation, villagers who leave their lands in fallow for the traditional fifteen or twenty years may lose them because "unexploited lands can be taken over by the State." So now villagers must keep cropping the land: exhaust it, or lose it. Of course, the law makes provision for villagers to register their lands, transferring their customary rights into colonial Roman Rights. But why would a village chief think of doing it? His ancestors have occupied and tilled this land for generations. The new law is written in French. Villagers do not speak French. They certainly do not read French. Rural villagers have never heard about the French law for registering land ownership. Who other than a venal civil servant who is able to speak and read French, could expect to understand the significance of these new laws written in French? Then, in return for a generous bribe, the government official will register a village mango grove and fishpond as the newly-acquired property of a wealthy city merchant.

Thus, Independence has become a plundering of lands by the new urban elites. Village soils are becoming exhausted: why should the peasant care for his land, if it can be taken from him so easily? Why should the peasant plant trees, when the trees he has planted are sold to city firewood merchants in exchange for bribes paid to officials in the Forestry Service? Some farmers wake up in the morning to discover that the yard outside their house now belongs to an army colonel. Why should old men struggle to protect their fields against erosion, when their young have emigrated to find a better life in the city? By the end of the twentieth century, half of Africa's population will live in the cities. And when the urban elites have destroyed their countryside communities, who will grow their food?

But all is not gloom. There are civil society organisations in Africa that promote justice and defend villagers. I have met development workers helping villagers to enrich the fields that farmers are still cropping. Projects run by agencies like Near East Foundation and SOS Sahel are organising land reclamation using simple technologies. To reduce soil erosion, farmers lay rows of rocks across their fields, following the contour lines. The contours are calculated using a long hosepipe full of water: so long as the water is not leaving the pipe, you are one the same contour level. Every ten yards or so, a new stone barrage is laid across the slope, each line of stones forming a barrier to block rainwater run-off. Slowing the flow of rainwater down the slope of a field, reduces soil movement. The rock-line creates a place for dirt to gather, where seeds germinate. It is fascinating to see how quickly vegetation takes root along the lines of rocks, where dirt and moisture accumulates, instead of being washed away. The roots of these fragile plants stabilize the soil, create humus, reduce erosion and protect crops against the effects of wind and rain.

The villagers showed me small dams of rock they had placed in a former gorge, gouged out by heavy rains that had eroded the land. Soil was now accumulating behind these barrages. After just five years - thanks to these rock barriers - a considerable surface of land had been recuperated and was planted. Whole hectares of land had been saved from erosion. I was standing on a millet field where there used to be thin air, in what had been a severely eroded gorge. The millet crop I was admiring, was growing on soil that did not exist five years earlier.

APRIL FOOL

It is extraordinary how people from abroad try to import (and impose) their own strange customs into other countries. Americans try to impose democracy and hamburgers – which many other cultures deplore. [Is anyone praising Taliban democracy in Afghanistan? What about democracy in Vietnam? How about cultures where people do not eat meat?] No nation is more obstinate in this regard than the British. Even after the end of empire, the British continued to create Clubs and Caledonian Societies, dancing reels and jigs across the globe and organising fancy dress balls. Deep in the Niger valley south of Timbuktu, an Englishman was working in nature conservation. One 1st of April, he decided to perpetrate an April Fool on his African colleagues.

To give him his due, he did make an effort to be un-British about it ..... but instead of trying to be African (and realizing that they don't know about April fools), he tried to be French. After all, they speak French around there. In all French colonies, French is the official language of the developing nation.

The Frogs also celebrate April 1st, but they call it "*poisson d'avril.*" It is mainly for children. Small kids race around the school playground, slapping friends on the back, leaving a paper cut-out in the shape of a fish stuck with tape on the shirt of the unwary friend. This is literally a "poisson," and primary school children find it hilarious.

Even French adults sometimes celebrate April 1st. On one famous occasion when Mr. Edward Heath (an honest and intelligent British Conservative prime minister) was about to take Britain into the European Common Market, French national radio included their news bulletin with the announcement that -- as a gesture to Franco-British solidarity -- the French government had decided that France would change to driving on the left side of the road. Within minutes the telephone lines were jammed with frantic calls. Panicky

Frenchmen blocked the switchboards of every ministry in Paris to complain—and national radio was forced to put out a second announcement explaining this was a "*poisson d'avril*."

To return to our Englishman, he decided to provide a *poisson d'avril* for each African colleague. On April 1st, each of them found an envelope with his name on it. Inside each envelope was a dried fish. No one enquired as to why they had received this small, unexpected gift. The Malians simply took their fish home in the evening and gave it to their wife to add flavour to their evening meal of rice and stew.

Not daunted, our Englishman tried again. The following year on April 1st there was a staff meeting. When the staff arrived, they found that all the conference room chairs had been replaced with children's bamboo chairs. Each person duly took his place. Their noses barely reached the level of the table-top, but nobody said anything. The staff found it difficult to write notes, scribbling at nose-level, but West Africans are easy-going people and they do not complain much. After a century of colonial domination, they are used to White Men doing strange things. At the end of the meeting, they all filed out, leaving the Englishman with his children's chairs.

Of course, it is possible that the Africans were getting the last laugh by pretending to ignore their boss's *poisson d'avril*. But Africans find foreigners so strange in their ways, that the abnormal seems normal: they will accept virtually anything incomprehensible, if it is desired by a White Man.

A White friend of my family recounted arriving home one evening to be told by his cook: "I done pluck de chickens like you say, and I done put them in de fridge. But they no happy." The foreigner looked into the fridge cabinet to find two very cold plucked chickens saying "cluck cluck" in a mournful voice. The White Man had said to pluck the chickens, but he not said to kill them: so the cook had not killed them. The cook found it very strange to be plucking a live chicken; but with Europeans, everything they do is strange. White men are perfectly capable of wanting live chickens plucked, using reasoning that only they can understand.

There are of course thousands of Africans who have travelled and lived in Europe. They pick up and understand European ways, because they want to learn them. Very few Europeans in Africa have a similar attitude. Most Americans in Africa live an American life, and Europeans are not much different: few bother to make deep African friendships. On the other hand, there are plenty of Malians in the country's capital city who know about *poisson d'avril* and April Fool's Day. A group of young Malians decided to play a prank on 1st April: they printed a poster announcing a free dance at the biggest hotel, the *Hotel de l'Amitié*. There is not normally much "*amitié*" (= friendship) there. It is more money, than matey. Everything in the hotel is so expensive, you have the impression that it costs money to breathe – and since it is all air-conditioned, perhaps that is true!

The youths called their fictitious ball "*Le Bal de l'Amitié*" and stuck up their notices around the student quarter. April 1st was the date announced for the ball. Several hundred young Malians duly arrived for the free dance, to the horror of the French expatriate management of the hotel. In common with all the best practical jokes, the "organisers" took care not to attend: they enjoyed their joke without needing to see the frustrated Malian students and their furious French "hosts."

TRADITIONAL MIDWIVES

On the edge of the desert, I visited a village where the newly trained midwife was quite useless. She had received her basic hygiene and birth-delivery training six months ago. Our Oxfam-funded project had provided her with the basic kit, including a clean rubber groundsheet for the mother to sit on (or lie on), alcohol and cotton wool, new razor blades for cutting the umbilical cord, soap and sponges, etc. Yet when I arrived with some colleagues, we found that her equipment was still as new as the day she had received it. We needed to know why.

As soon as we saw the 18-year-old midwife, we knew the answer. What mature village woman would allow an unmarried eighteen-year-old into her house while she was giving birth? Such an idea is unheard of! Her presence might even attract malicious jinns, and endanger the baby's life. I needed an answer to the question: "Since all the project protocols specified that midwives should be middle-aged women and mothers, how did such a youngster get chosen in the first place?"

Of course I intuited the answer: she is the daughter of the Head of the village. The women were not consulted and our field-staff did not feel strong enough to defy the choice of the village chief. He chose his daughter, and so his daughter was accepted for the training course.

In fact, all is not lost. The girl does not meet the three criteria of being married with children, being unlikely to leave the village, and having the respect of the older women. But her mother does. Her mother is one of the traditional village midwives, a natural leader among the women—and we found out quite quickly that she has learned a number of rules about hygiene from her newly trained daughter. It is of course the mother who should have gone for training .... but the old man wouldn't let her go.

The traditional birth attendant does not have a lot to do with the mechanics of birth. Mothers usually sit on a stool to give birth, or on a mat. The birth attendant is there to catch the baby, to wash it, oil it, and to burn certain herbs that protect the occupants of the birthing hut against evil. If there are two elderly women present, one may support the mother's back or hips, or press on the pelvic girdle to help with the birth.

The most important act of the birth attendant (both symbolically and hygienically) is cutting the umbilical cord. This is traditionally done with any convenient knife or stick, and neonatal tetanus is a common result. The introduction of the rule "a new razor blade for each new birth" has saved many little lives.

Other than that, the mature ladies who learned the trade from their mothers really have very little knowledge of what the birth process is, even though they have given birth themselves. What should they do if the placenta doesn't come out after the baby? A recent survey by Save the Children brings out a number of possible treatments, each more uncomfortable (and dangerous) than the last. Here are the options suggested by village women of the Sahel:

- stuff tobacco into the mother's nostrils
- make her smoke peppers (vicious red pili-pili peppers!)
- give her some cream to drink, then stick your fingers down her throat to make her vomit
- squeeze her belly with a cloth, or massage it (this is a better option than tobacco, provided it is done without violence)
- pick her up and drop her (good risk of haemorrhage)
- go to the local marabout priest-magician to obtain a "gris-gris" or "juju"

Two out of every three women consulted, said they believed in the vomiting solution – I suppose that vomiting violence could provoke muscle movements above the placenta. Only 6 out of

68 women suggested that the mother should go to see the trained midwife in the government clinic. Although pre-natal and post-natal care are widely accepted in the towns, in the villages there is still a long way to go.

The most picturesque "aid to birthing" was suggested by an elderly Fulani lady. "If the woman's labour is too long, the husband may soak the belt of his trousers in water and then give the water to his wife to drink. But this solution is also very dangerous. If the baby is not the child of the husband, then the wife will die at once." This test of paternity is simple to carry out and apparently radical in its effect. I recommend it to all doubting husbands.

HOT RICE AND WELL WATER

I wish this rice was not quite so hot. Oh, and the weather too: 105 degrees Fahrenheit in the shade is painful, but so is the rice. Squatting around the common eating bowl, taking food with my right hand, I realize that my fingers are more sensitive than anyone else's. I am getting less food than the villagers, even though I am just as hungry as they are. My very hunger gives me the impression that Karim next to me is eating twice as much as I am. Karim is a friend and colleague, working for a Malian NGO.

I remember Ced, a young volunteer working with Fulani nomads, who nearly died of starvation. Now Ced Hesse is an experienced member of the Oxfam staff and publishes articles in their very interesting fieldworkers' magazine Baobab. As a volunteer, Ced's Fulani rice was always served too hot. Tears of pain streamed down his cheeks as he balled up some rice in the palm of his hand, trying to ignore the hideous fire torture scorching the tips of his fingers. The rice and sauce were tasty enough: Ced simply couldn't grab enough of it. By the time his stinging fingers had fed three balls of rice into his mouth, the dish was empty. The nomads thought he had a small appetite and teased him about it. All through the meal, his Fulani nomad friends encouraged him to "eat more, your hand is too slow," not realizing that they were adding to the finger-scalding pain by emphasizing his hunger before the meal and his inevitable emptiness after it.... not to mention his feeling of inadequacy in not being able to handle the anthropological experience of eating out of the communal dish. In compensation, I suppose the teasing probably enriched his Fulani vocabulary.

My fingers are stinging too, but the pain is bearable. Worse is the inelegance of my movements. At least they seem this way to me. I grab the hot rice with some sauce. Then I squeeze it into a ball, which is somehow less round than that of my neighbours. Raising my hand, I stuff the ball into my mouth:

Karim seems able to just pop it in. Some of my rice-ball always stays behind, whereas Karim's rice-ball seems to fit his mouth. Then there is the mess. Karim is black, and his right hand glistens with the grease. My white fingers look like a muck heap. The bits of green and red and orange show that the sauce is rich in tasty vegetables, but they stand out on my fingers like a broken kaleidoscope. While the others are licking the juice off their fingers and going in for another handful, I am desperately sucking minute pieces of carrot and tomato off my fingers so that my hands don't look so messy. Quite absurd, of course. No one cares except me. The others simply think that this is a good meal, and they are vaguely pleased that the White man is sharing their food without any hint of disdain.

My hunger is justified since I have been working hard all morning in a seminar on "village organisation and partnership." The organisers and animators of this village jamboree are young field workers who have been given full responsibility, and who are doing famously. There are sixty people participating (only about twenty of them are women, but that is not a bad ratio) from some thirty village groups: agricultural cooperatives, women's garden groups, some semi-nomadic beef-fattening associations, men's beekeeping groups, female soap making teams, a couple of mixed village savings banks with a scattering of field workers from five different local voluntary organisations. This is civil society at work, and working with success.

Logistics are easy. People will sleep anywhere so long as they have a woven grass mat—and so will I. It is more comfortable on a mattress, I will admit, but a woven mat is fine. Being willing and able to sleep anywhere, is a part of my professional skill set. Sometimes in the villages when the houses are stifling hot at night, I choose to sleep outside on a hard wooden table. Tasty food is always prepared by the host women, and this week we have plenty of wonderful fresh fish from the fishing cooperative coming straight out of the Niger River just fifty meters from our meeting hall. Oranges are in plentiful supply at this time of year, so I never have to go thirsty.

My only real fear in the bush is raw milk. Unless it is boiled, I say "No, but thank you." TB can be cured these days, but I absolutely do not want to catch brucellosis, also known as "undulating fever," as "Fulani fever," or as "abortion fever." On one occasion I was handed a very dirty-looking calebasse (a round gourd split in two, making two small basins) filled with fresh milk. I wet my lips, and my host objected that I had not really tasted his milk. Even in primary school, I did not care for drinking milk. However, needs must in development anthropology, so I lifted the grubby calebasse to my lips for a second time, giving myself a white milky moustache up to my nose and assured my generous Fulani host that his milk was delicious: he whose home I was visiting, had to be content with that. At least I did not catch brucellosis!

Water also carries infections. I try to drink water only from a well. Once the water has left the wellhead, it usually travels across the village in an open bucket and festers inside the family clay water jar. I try not to drink water from any communal jar; or, if there is no alternative, I try to drink as little as possible. I cannot tell how often they disinfect the clay jar.... but I can guess. If – instead of drawing water from a well - the water comes out of the river, then diarrhoea is a certainty with guinea worm and bilharzia as distinct possibilities. I'd rather go thirsty.

These community debates are thirsty work. The villagers are exchanging experiences about their economic self-organisation. Are their neighbours doing better than they are? How are the profits in their mixed cooperative? Do women earn bigger revenues when they have their own autonomous structures, separate from men? The secret for getting peasants to exchange their experiences, is to ensure that the non-villagers shut up! This morning was largely wasted, because government civil servants never stopped talking and the meeting's chairman never called upon even one woman to give her opinion. Karim and I asked several questions, trying to steer the debate towards the women, but each time a technician barged in with his own answers.

Before lunch, we had a quick meeting with the organisers, to see whether the others had noticed what I had noticed. Karim was made chairman for this afternoon and we have agreed that anyone who has been to school will be forbidden to speak. I know that this will make the government forester really unhappy since he doesn't know how to shut his mouth (even when it is full of food). And the local chief of the government extension service will be annoyed, since he considers himself the most senior person in the room. Government people are hierarchical and they consider that speaking rights are in direct proportion to seniority, salary, and graft (whether they have anything interesting to say is irrelevant).

But we are non-governmental (and sometimes anti-governmental). We and the organisers have agreed that since the peasants are chief guests, the most senior participant has to be the oldest villager. So, this afternoon the older peasants will be given priority, then the women, and then the younger men. Officials therefore come last. My function this afternoon is to support Karim as he attempts to keep the technicians quiet, so that the villagers can tell us what they really want and need. The real experts are the villagers. We want to learn about their practical experiences. The local chief of the extension service and the local forester may even learn something this afternoon, if they don't fall asleep from stuffing their mouths with too much rice.

CROCODILES IN THE BATHTUB

I called into the office of an eccentric development manager I know. I do not mean this in any critical way. He's a very professional manager is Bob, an experienced village development worker with lots of ideas and a vision of a world with less poverty. He works for a very respectable bilateral agency, but he's a bit unusual. He's sort of eccentric in a very British sort of way: not surprising really, since he happens to be British. On this occasion I called in to talk about soil erosion and desertification. Instead of which we talked about crocodiles.

At the entrance to Bob's office, I braved the bouncers in baseball caps, and signed a register in which I had to write the name of my father and mother, and even my own name. This rite of passage entitled me to receive a Visitor's Pass, and to enter the Hallowed Portals of a shabby building dominated by the rattle of a noisy air conditioner. I hunted around for a chair that didn't have papers or computer tapes on it. Bob offered me some tea, and produced instead a spoonful of dust, which proved to be something like lemon-flavored. Not natural lemons: rather a European laboratory lemon essence sort of flavour. The bottle said "Nestea," and claimed to come from Germany. They don't grow lemons in Germany. Anyway, I sat down with my mug of lemon essence, and found myself staring into a trash-bin full of baby crocodiles.

"Rather convenient having a metal waste-paper basket, don't you think?"

Only a Brit could open a conversation so nonchalantly. It seemed normal to him, having a trash-bin full of reptiles. Standard office furniture. But I was sitting open-mouthed, my eyes popping out in the direction of the crocs..... and they were staring back with equal amazement.

"A fisherman offered them to me. He dropped his net down into the Niger River near the Guinea frontier, and up came some baby crocodiles. They should have been hiding in their mother's mouth; but she perhaps had her mouth full. When baby crocs feel fear, they hide in their mother's mouth; but I suppose they did not hear the net sink through the water as the lead weights closed around them."

The fisherman has to make a living. He had no Nile perch for market that evening, but he did have some Nile crocodiles. Who might buy baby crocs? A foreigner, naturally. So he spoke to Bob, and Bob bought them. Big fish are sold by weight in the market; small fish are sold one-by-one, or in heaps. These were small crocodiles, about 18 inches long. Bob bought his four crocs as a heap for $50, took them to the office in a cardboard box, and decided to store them in water in the trashcan.

"But you cannot keep them in a trashcan!" Of course not: that evening Bob took them home, and put them in his bathtub. He made a special point of not telling his wife. He says she saw them while she was under the shower, and he claims that her scream of fright kept him chuckling all night. That must be the "British sense of humour." I asked his wife if it was true. She said, "Of course not! The children couldn't wait to tell me as soon as I got home. You don't want to believe everything Bob says." Now where have I heard a woman's voice saying something like that before?

The crocodiles seemed to do very well in Bob's bathtub. I am one of the friends who have been helping with cockroaches. Baby crocodiles love cockroaches. Never in West Africa did it ever occur to me that the day would arrive when I was short of cockroaches! In Sierra Leone they ate most of our library. In Nigeria we never dared put a hand into the drawer to take something, without first opening the drawer to let the roaches jump out. Now, in the dry Sahe,l I wish there were a few more big cockroaches around to feed Bob's baby crocodiles. I can hear some of my female readers thinking, "Do they really eat such horrible food?"

We popped one juicy cockroach into the bath, and sat down to watch. Our roach was just at the perfect stage: incapacitated, stunned, but with one leg still flickering to show that it was fresh. This lovely morsel floated on its back towards the baby crocs, who ignored it. The plug hole of the bath was stopped with a piece of cloth, allowing the water to leak slowly. This created a slight current in the bath, the water very gently rotating around the bath. The roach floated slowly round again, while the four crocodiles remained absolutely motionless. Even their hooded eyes didn't move. Crocs have lids top and bottom: you would think this might make them blink twice as much as we do, but not a bit of it. They only blink when they want to, or if you drop a gob of saliva straight down onto their eyeball. And then they only move their eyelids. Nothing else.

So, there they were, sitting one on top of the other under the water, their little beady nostrils just at the surface of the bath water. Crocodiles are reptiles that co-existed with dinosaurs 200 million years ago, and outlived them. Survival has taught crocs to conserve energy by moving as little as possible. The cockroach floated on gently, one leg flickering. I glanced at my wrist. We had been watching that damned roach for 6 minutes! And still not one eyelid .... SNAP as quick as lightning, one snout had flashed. The only way I could tell which of the four crocodiles had moved, was that the cockroach's whiskers were visible between the teeth of one snout.

Every two or three days, Bob let out the water, rinsed around the tub with the flexible shower, and put in more water. Then one day the phone rang while the bath tap was running. When Bob came back from the phone, the water had reached nearly to the top of the bath.... and only three baby crocs were still swimming around the bath. Oh dear! What had happened to the fourth croc?

We called my friend Dr. Keita, the veterinarian at Bamako zoo, to ask his advice for Bob. "These are Nile crocodiles," confirmed Keita, picking up the specimen we had carried with us. "They can move very fast. And they can climb trees (which Bob doubted). If the baby crocodile escaped, then he probably made his way back to the river, but he will not survive in the river. Either a big fish will swallow him, or else another crocodile. The biggest danger for baby crocodiles is bigger hungry crocodiles. This is why they stay near their mother, and swim into her mouth if there is danger."

We decided that the crocodile was gone. If it can run fast, it might have got out into the garden and the river is not far from Bob's house. We went around the garden with little conviction. Bob was hunting under the bushes. I confess that I was looking up in the trees, seeking a crocodile disguised as a branch. The missing crocodile turned up three days later. In the house. He (or is it a she? I haven't got far into crocodiles yet) was found flapping around the son's bedroom and needing water.

Bob picked it up. The slender tail flashed, the head squirmed, the vicious mouth let out an indignant squeak and one razor-sharp tooth made a red stripe down Bob's finger. It looked like a paper cut: the cut you can sometimes get from the sharp edge of a paper envelope. A reminder that Nile crocodiles - even tiny crocs no bigger than a lizard - are dangerous, wild creatures.

This specimen didn't seem very speedy on its legs. Climb trees? When you actually get down to the geometry of a baby Nile crocodile, Bob is quite right - its legs are not long enough for climbing trees. Even running seemed to be difficult, as it was using its tail muscles to leap along, or "bump along" the tiled floors. But back in the water it moved quick as an eel. And then sat motionless on top of its brothers .... or sisters.

So, Bob is back with his four crocs, and they are still in the bathtub. But not for long. Bob's wife has delivered an ultimatum to get them out of the house and into a pond. Now I have to confess that I'm rather hooked on these babies, so I'm helping him design the new crocodile house in the garden. It needs water for swimming and a sand bank for them to rest on. I wonder, does it need a roof? Or a food larder? I'll keep you posted on their progress.

CROCODILES IN THE GARDEN

The crocs have been moved out of the bathtub. These are my friend Bob's baby crocodiles, which he kept in the bathtub at home: partly because it was a convenient wet place.... but mainly because it made for such excellent dinner conversation. Each time a lady guest went to look at herself in the bathroom mirror, she stood up relieved, and found herself staring into the beady eyes of four Nile crocodiles. Four long, thin, black babies, only the size of a lizard but obviously crocodiles. Male guests who knew about the crocs would pause in their conversations, their beer glasses suspended in the sultry night air, listening for a shriek, followed by the sound of the flush, and the patter of agitated female shoes along the cement corridor, and the excited question, "Do you know you have crocodiles in the bathroom?" Yes, Madam, we all know and we all heard you scream.

The gag worked just as well with African women as with foreigners: women will be women, you know. But like all harmless games played by us immature men, THEY put a stop to it. Bob's wife said "them or me." After careful consideration and a mature discussion with his male friends, Bob summoned a mason to build a pond in the garden. Best not to call his wife's bluff.

The crocodile "cage" we designed consists of a cement basin for swimming, and a sandpit for sunning. That is all crocodiles do, apart from eating. And because they do so little, they eat seldom and live long. There is a theory that the sand might also be good for laying eggs, but I guess these crocs will need a bigger cage before they reach the stage of puberty. They have already doubled in size since June when Bob bought them from a Niger River fisherman.

The baby crocs probably do not know yet whether they are male or female. Nor does Bob. They are siblings. Incest is a forgone conclusion. Apparently the males have a double-barrelled penis, which should give them an adolescent advantage over ordinary mortals like .... well .... like me for example.

We didn't think the croc cage would need a roof: just some shade. Over the top of the enclosure there is a straw roof swaying on a typically African stick arrangement. One of the workmen has thoughtfully suspended from this roof the ear of a warthog. The gentleman who did this is a Malinké, and he won't (or can't) explain the value of this dried grey cone of leather. Obviously, it is beneficial, and presumably it protects somebody from something. Since the workman lives nearby, it probably protects him. I dare say there is not a house in the whole of Africa which doesn't have a bag of grasses, or a bunch of feathers, or an oryx horn full of pepper seeds, or a lucky leather pouch with magic words hanging or buried somewhere around it. Well, Bob has a warthog's ear to bring him luck.

I was invited round for the ceremonial transfer of the crocs. Jeanne sat under the mango tree with Bob's wife and feigned indifference. The kids were enthusiastic participants, and invented their own version of mother-baiting, by deliberately carrying the baby crocodiles as close to the mothers as possible. The process was extremely simple: 1) remove plug from bathtub; 2) pick up each croc with a cloth to avoid razor-sharp teeth; 3) carry same to new croc-tub and deposit gently into dry cement pond; 4) bring hose pipe; 5) turn on water to applause of assembled children and discrete giggles of supposedly snooty mothers; 6) watch babies swim around in their new home, propping themselves up on their front feet to hold their long snouts under the water fall. Their eyelids closed top and bottom (they have two which conveniently meet in the middle of their protruding eyeball), they were enjoying the flow of water spraying over and around them. It obviously gave them immense pleasure. I have seen big crocs do the same beneath the magnificent Murchison Falls in northern Uganda, said to be the most powerful water fall in the world, where the Nile River forces its way through a 7-meter gap in the rocks, and crashes down the 43 metre drop (140 feet) with dramatic effect. I remember standing on rocks at the top of the Kabalega (or Murchison) Falls, and being cooled by the spray erupting from 140 feet below. When you approach the falls in a boat down below, you can see the crocodiles enjoying the spray, just as Bob's baby crocs were enjoying the hose pipe: the primeval instinct repeating itself across the continent and across the millennia.

Crocodiles have lived on Earth since the dinosaurs, 200 million years ago. Dinosaurs were wiped out 66 million years ago, but the crocodiles survived. They are the ultimate ancestor figures. Most people believe that crocodiles bring luck. The city of Bamako (capital of Mali) beside the Niger River actually gets its name from the Nile crocodiles that live nearby. Bamako means "old crocodile" or "place of crocodiles" ('bama' = croc but 'ko' has more than one possible meaning). The armorial crest of Bamako city shows three crocodiles with curved tails, sitting one above the other. Bob's baby crocs often spend their day like that, piled up in a favourite corner of their pond, one on top of the other with their snouts just breaking the surface of the water. Motionless.

If Bob strokes them, their formation breaks up in panic. Two baby crocs seem to like being stroked on the back of the neck with a brave finger, while the other two flee to the far side of the pool, paddling with their four feet. Suddenly they pull their feet into the side of their bodies to form two elegant streamlined torpedoes with powerful propelling tails. Clumsy on land they may be: but in the water they are magnificent!

They are getting fat, these babies. Their diet of meat, minnows, and cockroaches is doing wonders. At the end of the wet season, Bob saved a lot of money by picking up small frogs and toads in the grass, and tossing them into the crocodile pool. Poor little frogs! Lucky little crocodiles! Actually, the crocs are supposed to prefer dead meat to fresh frogs. Traditionally crocodiles "hang" their game like Scotsmen. I remember the wonderful scene in James Clavell's book (and film) Shogun, where the English hero hung his pigeon to get ripe, while the bemused Japanese made larger and larger detours to avoid the stench of his rotting bird. Any well brought up crocodile creates a larder - often under the roots of a riverside tree – where it stores fresh meat until it stinks enough to make a decent meal.

For people living in Africa, it is recommended to study a crocodile's eating habits, just in case. The crocodile traditionally catches his prey (for example an unwary Green Monkey needing a quick drink, who wasn't watching the floating logs; or a thirsty antelope whose delicate hooves strayed too far into the edge of the water) and drags the victim under the water. The croc turns his prey over and over until resistance is gone: which means that crocodile victims usually die from drowning. Once the prey has become meat, the crocodile swims to a hole dug into the side of the riverbank and pushes its next meal up above the water line. The meat is left to rot ... or ripen according to your culinary preferences.

Our local tailor Mr. Naré claims to know all about crocodiles. His grandfather was a famous crocodile hunter, back in the days when hunters used spears and crocs were plentiful. He was explaining to me that a small piece of crocodile skin is often the essential wrapping for powerful traditional amulets to wear around your arm, neck or waist. He looked at Bob's brood, and nodded sagely. "They have already started eating stones" he said. We raised our eyebrows, nodded to the warthog's ear on the other side of the pool, and waited for more.

"You know of course that crocodiles have a separate pouch near their stomach in which they keep stones?" It must be their appendix, said Bob irrelevantly. "And when you kill a crocodile," the tailor continued, pretending (quite rightly, I thought) that Bob wasn't there, "You can tell the age of the crocodile from the number of stones in this pouch. Because crocodiles eat one stone every year."

So that is how crocodiles keep track of the calendar? "Happy birthday, dear cousin crocodile.... and this year we have chosen a beautiful piece of granite for your birthday tea." I am sceptical. Mr. Naré is an entertaining old rogue, often drunk, and I take his tale of stones with a pinch of salt. It sounds to me like is a shaggy crocodile story. Crocodiles can live past the age of 60. Sixty stones would make a very heavy load for an old croc to carry in his belly. Surely he would sink?

Could Stone Henge possibly be the remnants of a birthday party from a giant race of extra-terrestrial crocodiles? I think there must be another explanation, both for Stone Henge, and for the crocodiles' strange propensity for eating stones. If you know any crocodiles, could you show them this article and ask for an explanation?

The Miami Science Museum reports that stones help crocodiles swim by providing ballast weight, and that they also help crushing food. Such rocks are known as gastroliths. "Rocks in a crocodile's stomach help crush and grate food. Rock swallowing is especially beneficial for crocodiles who eat whole prey, particularly animals with shells and tough bones," says the museum website.

Bob feeds his crocs a diet of meat pieces, and occasional toads that do not require crushing with gastroliths. Nor will they ever be taught by their mothers how to keep a well-stocked larder. They will never learn that it is advisable to keep the rotting monkey separate from the fresh antelope on the other side of the larder. They will never even taste a ripe Green Monkey. I cannot see how they'll taste a human, unless one manages to take a bite out of Bob's finger. Their destiny is never to catch anything bigger than a frog: and even that is cheating, since the frogs cannot jump out of the walled enclosure. Here is a typical case of rural outmigration producing spoiled urban youths who have never had to work for their food. They just lie there waiting for pieces of rotten meat to fall into their rotten mouths. Typical lay-about adolescents. No wonder they are getting fat. How soon before they spend all day listening to Bob Marley or some Rap band?

PS: This story was written 30 years ago. Those small crocs have grown and reproduced. Incest has happened! The biggest in the family is now bigger and heavier than a human. The crocodiles live in a much-enlarged enclosure in Barou Tall's farm, on the banks of the Niger River in Magnanbougou. The population now numbers in the dozens. My wife keeps telling me she wants a crocodile skin handbag.

LIFE AND TIMES OF A GIANT TORTOISE

We have just acquired a giant tortoise. One of my delights relates to the fact that tortoises are ancestral figures for the Dogon people of Mali, who are reputed for their ancient culture and their grasp of humanity. Dogon wood carvings show that they revere tortoises and crocodiles among their Ancestors. We know that tortoises and crocodiles co-existed with dinosaurs, making them among the most ancient precursors of humankind. Dogons say they are descended from mythical twins called *nummo*, who may be represented as human from the waist up and serpent-like below. Somehow or other, the Dogons have always known about evolution (long before Charles Darwin) and the fact that we humans are descended from earlier forms of aquatic and amphibian life. Serpents, tortoises and crocodiles feature as ancestral figures alongside the *nummo* in all the Dogon creation myth carvings I have seen.

We already own some crocodiles and, like me, Old Brother was delighted when we added a giant tortoise to his farm livestock. I thought these prehistoric creatures (the most ancient of all the reptiles) were quite slow in their movements, but ours seems to be an active adolescent. Each time I find him near the gate, I pick up his 15 kilos and carry him back to the grass. I don't say he's actually an athlete: no one is likely to be a sprinter when they have to carry a bungalow on their back. His four legs protrude at quite the wrong angles for athletics, forcing him to hobble on the outside of his two toes, rather than on the foot pads. He is not a sprinter.

Furthermore, our tortoise refuses to move any faster than his suspicious head can see. There is the head swaying to and fro like some prehensile robotic arm, withdrawing into its shell at the first hint of the dog. In the hope of provoking an incident, I tried placing the dog's food on top of the tortoise. He showed no interest whatever in rice, soggy bread, meat or milk. Clearly our tortoise is a strict vegetarian and herbivore, just like our book says he is.

I say "he" but really have no idea what sex "he" is. The book says that I cannot look and see. Apparently even tortoises don't know what sex they are. They cannot tell each other apart until the female comes into heat. Then the males suddenly discover that they are males! They smell the female, know that she is different from them, feel their hormones bubble, and lumber up to see what they can do about it.

Bamako zoo has tortoises. They are there, down in the field with the broken fence where a family of gazelles and two Indian buffaloes graze cabbage leaves and old plastic bags. The zoo tortoises look ancient. No athletic adolescents in the zoo: they barely move at all. In fact, you can only spot them if you assume that all grey rocks are really tortoises and make a careful study of the rocks. Maybe there are also some babies disguised as stones.

Now these are real giant land tortoises, even if none of them approaches the half-ton monsters you can find in the Seychelles. The race has been here for about 40 million years, so they have been successful survivors. Their eggs have soft shells, which they bury in the warm earth where they leave them to hatch. Having just had a bad night with our no. 3 infant, I feel pangs of jealousy towards the tortoise. I feel a guilty attraction to the idea of burying my offspring and leaving it to survive in a warm, natural habitat. How does one apply to be reincarnated as a giant egg-burying tortoise?

I am not being serious. It is the reproducing thing that is the greatest obstacle to my reincarnation plans. For I have recently been witness to one of the great dramas of zoo living, one of the truly touching scenes of the African life cycle: I have watched a pair of giant tortoises making love (presumably discovering in the process which of them is male and which is female).

First of all, I heard them. We were walking out of the zoo when I heard the noise of sawing wood. I thought it strange. It is true that the military authorities are ruthless about cutting down trees of beauty, selling the wood for profit, and replacing them with hideous stretches of black, heat-storing tarmac. But surely not in the zoo? Anyhow, are there any trees left in the zoo to saw?

I have watched village lumberjacks sawing tree trunks of red cailcedrat wood (kaya senegalensis or Gambian mahogany) into planks, using huge 2-meter hand saws operated by two men. The top man stands on the massive trunk, which is balanced on a trestle six feet up in the air. His mate stays on the ground below and they operate the saw by heaving it up and down using both hands and shoulders, while dark red sawdust floats around them and clings to their knotted black muscles. That is what I was looking for when I was stopped by the sound of sawing. No trees. No lumberjacks. I leaned over the parapet, through the broken fence. And there below me were the tortoises.

It was Mr. Tortoise who was making the noise, heaving clumsily atop a grey rock which turned out on further inspection to be a female tortoise (I suppose: but if the boyfriend cannot normally recognize a girl, and if this was his first time, and in a world where gay marriage is increasingly accepted, then there is considerable room for speculation. Maybe boys and girls don't make a sawing noise....).

I don't know whether you have looked under a tortoise? Below the great dome of the back lies another rigid shell, a heavy flat one. So even once you have heaved yourself onto Madame, it isn't very intimate. This is what it must have been like for medieval knights in full armour trying to rape a woman wearing an iron chastity belt.

The tortoise mating process is not cheek-to-cheek, more like chassis-to-chassis. Perched up there, the male tortoise looked like a clumsy helicopter in the process of taking off. The joystick had just been pulled back for take off in a manner of speaking. The position of the tortoise wouldn't suit me. Body contact is minimal. In fact it must be limited to ..... I am having second thoughts about reincarnation.

It is just as well, given the difficult circumstances of the mating, that tortoise sperms can swim around inside the female for several years without losing their reproductive potency. Since the tortoise has been around for so much longer than the rest of us, this spermal particularity must be a major part of the secret for survival of the species.

Despite the ungainly appearance of the coupling, the effort is not in the heaving to get up there on top but in the heaving and knocking thereafter. Aah- haah, ugh- hugh, aah- haah, ugh- hugh...... the sawing noise went on and on and on. Chet aged 14 thought it was fascinating. Leïla is 11 and giggled. Jimmy aged 3 was bored, and he won. The mating process can apparently last up to 5 hours. The pair we were watching were pretty big. They may be nearly 100 years old. At that age of course, five hours doesn't seem very long, but it was too much for Jimmy. I must apologize to my readers for not having stayed so as to be able to tell you how it ended.

EUROPEANS SHOWING BAD MANNERS

Fatou was an economics graduate student in France. She is lively and intelligent, enjoys nightclubs, and has lots of friends. In the student flat above hers there were three French boys, frequent visitors to share a coffee with Fatou and her two Malian flat-mates Fanta and Ami. Their relationship was not "boyfriend / girlfriend," rather "good friend" and occasionally "let's go off as a group of friends for a Saturday evening" to a student dance. One evening over coffee, François threw out the idea, "Why don't we go to Mali for our holidays?" Gilles and Jean-Charles thought it was a great idea. Later in the year they set it up as a definite fixture with Fanta.... Gilles pushed the arrangements, especially since by now Gilles and Fanta had changed their relationship to the status of "copain-copine" - what the French call "Saturday-night athletes."

By the time the summer holidays arrived, Fanta had broken off with Gilles, but the boys came anyway. Fanta had backed off and Ami was away for the weekend, so Fatou and her brother met them at the airport. Fatou was now - by elimination - the hostess. She told her parents there would be three visitors: her brother moved to the neighbour's house, leaving his room for the three Frenchmen. Such is African hospitality. No one is ever refused. We share what we have, and everyone is happy to share it. As the proverb goes, "If there is food for one, there is food for ten" because the plate of food will be shared with all ten people, however small the share for each.

In fact you cannot go past a group of Malians who are eating, without wishing them "bon appétit." It would be very rude not to greet them, and they will immediately invite you to eat with them. Twenty times a week, I go through this simple ritual with colleagues taking their midday break, or with workmen eating in the street, or with Old Brother's farm workers at the gate. Sometimes I am hailed by Old Brother himself, taking a late breakfast in the shade of a mango tree. In the case of Old Brother, who is a very dominant personality and Head of our

"Family," it is almost impossible to refuse. So, I walk across to the mango tree and take a seat for two minutes, sip some tea, and listen to his advice while I gaze at the brown knobby knees sticking out incongruously from beneath his embroidered nightshirt. Having fulfilled my obligations of politeness, I then ask permission to "take the road" and make my escape.

In the case of the group of street workmen squatting with their right hands plunged into the communal greasy rice bowl, I nearly always decline the invitation to eat. Now, it is not polite to refuse unless you use the one acceptable excuse. You smile and thank them and say, "I am full up," or more correctly in West Africa, "I am fed up." Or better still: "I am fed up and agreeably drunk." Among the expressions to learn from your tourist phrasebook, "I am fed up" is among the most essential. Unless of course you have a liking for greasy rice. And workmen.

If I actually know the workmen, then I sometimes stop from politeness, even if my belly is really full. The etiquette requires you to bend your knees, and the squatting workmen will shuffle around to make a space for you around the bowl. Rinsing the fingers of the right hand is a necessary ritual (although it does not always guarantee hygiene, if you are the last person to get the bowl of dirty hand-washing water). And then you take a small grab of rice, knead it into a ball, and pop it into your mouth. Do this twice and honour is satisfied. The workmen are delighted that you have not taken too much of their lunch, and even more delighted that a strange White man has respected etiquette and shared symbolically their communal meal. Often, they're also delighted that I have made the gesture of eating with my fingers.

Someone should have explained these customs to the three French students, François, Gilles, and Jean-Claude. They told Fatou's family that eating with their fingers was a dirty habit. That was only a minor insult compared with the rest of their behaviour. We left Fatou and her brother at the airport waiting for their guests to come out through customs. When they appeared with their suitcases, they were all wearing pyjamas. Fatou was amazed. "Why ever are you wearing pyjamas?" she gasped. Gilles roared with laughter: "Oh, we thought we would give you a nice surprise. In Africa everyone wears pyjamas."

Fatou is a lady who knows what she wants and she'll not take nonsense from undergraduates. "Well, you are wrong," she snapped: "I shall not introduce to my parents unless you are wearing proper clothes. Either you dress properly or I shall take you to a hotel." So there and then, in the airport car park, the three boys had to change back into the blue jeans and T-shirts which they had removed in the plane. Their first joke had misfired.

Like all Frogs, these students were at pains to explain that French food is the best in the world. Mostly they timed their explanations for meal-times. Fatou eats student fare in France. In Africa, she is African and she eats what her family eats. She comes from a solid middle-class urban family. Her father is a merchant, her mother is a teacher, and they eat very well. I often eat there myself. If Jeanne is working in the bush, I generally honour my Malian friends by eating with them. Fatou's family is a favourite because they are good company, and because the food is always good. They hardly ever eat millet stodge (sorry, millet toh), so I usually expect rice for lunch. The Frogs found rice boring and told Fatou's mother that she ought to have a more varied diet. Charming! They suggested steak and fries. So, steak and fried potatoes were provided, but the boys didn't like the fries because they were cold. Well, in West Africa, food is eaten cool.

Ami returned to Bamako. The arrangement now was that they would eat midday chez Fatou and evenings chez Ami. This let Fatou's family off part of the overcrowding, as well as spreading the cost of feeding three large men. The Frogs did not offer to pay anything; but they are students, so no one minded that. The French students were glad to see Ami and glad for a change of restaurant. But they told Ami that she didn't vary her diet enough either, and they didn't like her fries any better than Fatou's.

One evening, when they had been there for almost two weeks, Jean-Claude confided to Ami and Fatou. "Are there nightclubs here where you can meet different types of people?" Ami laughed and murmured in Bambara to Fatou, "They are tired of us, they want a wider choice of girlfriends!" Then, "Yes, of course. What sort of people would you like to meet?"

Jean-Claude beamed, "Well, we would like to go to a night-club where we can meet some ...... French people." So, Ami and Fatou took them to a French hotel where they could meet some French people. The three of them sat in a corner and got tipsy, while their hostesses danced alone. What the Frogs really wanted, apparently, was alcohol. Malians are Muslims and few Malians drink alcohol.

The greatest transgression concerned their clothing. François likes to wear shorts. That is all right inside the house, but it doesn't go down very well in the city streets, in an African society where people dress for prestige and in a Muslim society where people dress for modesty. Shorts to a Malian look like underclothes – they breach both African and Muslim custom.

Fatou's mother is very religious, so shorts don't go down very well even indoors. François didn't like to be asked to wear trousers - but whose house is it anyhow? Gilles liked to wear shorts without any shirt, which is tantamount to being naked and the older women were profoundly shocked by his disregard of what they considered "decent" clothing. The greatest booby belonged to Jean-Claude. He felt ill, and lay in bed with a fever and a running stomach. He felt hot, and lay on his bed undressed. Then he felt nature call, and so he got off the bed and crossed the courtyard heading for the toilets. Naked. Fatou's mother coming out of her room was confronted with a full profile (and I mean FULL) of a lithe young Frenchman. Uncircumcised at that! She fled back into her room.

Fatou was at the end of her tether. I was summoned to the house for a family council and then delegated to tell the French boys that it was time for them to visit other parts of Africa. They left the next day. We hope they will not come back.

IN AFRICA WE TAKE TIME TO SAY HELLO

Coming back to Europe from Mali, the first thing that strikes me is the hurry. Why are we all rushing around? In Africa it is too hot to rush around: but we also take the time to live, to be social human beings. For example, we would never forget to say "hallo." In Africa, if you do not start with the correct greetings, you are impolite.

A British "expert" once led me into a ministerial office and asked the ladies of the secretariat: "Excuse me, is the Assistant Director in?" There was silence. Irritated, he repeated his question. More silence. He looked foolish. Meanwhile, I was making the rounds of the typewriters, shaking hands with the elegantly dressed ladies as West African custom demands. Seeing the discomfiture of my colleague, I put on my warmest "Saying hallo" smile and asked when the Assistant Director would be available. At once the secretaries were smiling and helpful. My British colleague looked sickly.

I drove a French friend up to the appalling general hospital on the hill, where there is no anaesthetic, no alcohol, no bandages, no plaster. Chantal needed to collect her husband's back X-Ray. "I want to see Dr. Coulibaly," said Chantal through the reception window. The plump girl slouched over a desk didn't even raise her head. Chantal was instantly exasperated: I could see that she was about to stamp her elegant foot like a Parisian lady. I pushed her gently onto a bench. "Watch how it's done." A young man in a white coat came walking along the path, and I hailed him. I shook his hand, smiled, asked the health of his family. And I then I asked him the name of the girl behind the window. She was called Fanta.

"Fanta, good morning, my Sister," I called through the window. The plump girl slouched over her desk looked up immediately, a broad smile on her face. We shook hands through the metal grill covering the window. Three minutes later I had Chantal sitting in Dr. Coulibaly's air-conditioned office, where we met all the other people who know that it is important to take the time to say "hallo."

Some people take greetings to an extreme. Among the Dogons of Mali and Burkina, it is thought unlucky to be the first to stop greeting your friends. Two farmers walking along will call out greetings across the field, stop for an instant, then carry on their way still greeting the person who has passed them. You meet old men and old women still muttering "Seywa, Seywa" to someone who has disappeared over the horizon. Then you greet them, and they start all over again. That is their custom, and a visitor should respect it.

This is especially important for a community development worker. If you miss out the greetings in a village, you might as well not come back! Courtesy requires a visit to the Head of the village before you meet with anyone else. Arriving at his front opening (he doesn't need a gate), you announce your arrival. Reaching the door of his house across the yard, you remove your shoes to show respect, seize the hand of the old chief, and pump it up and down while massaging his scraggy forearm with your left hand. The more you squeeze with your left hand, the more respect you are showing. Smart towny types do not remove their shoes. They shake hands with their right hand, keeping their left hand in their pocket. Townys are not successful in establishing warm relations with a village community. How a person greets the village chief is a good quick guide to how he or she will succeed in animating village development activities.

After the initial greetings, I usually present three cola nuts to the village chief as a sign of respect. I buy a bunch of cola nuts in Bamako before I set off up-country, because most cola nuts in Mali are imported from the forest lands of Guinea further south. I make sure I choose big nuts with unblemished skins, twice as many red nuts as white nuts, and I keep them moist, wrapped in leaves from the cola tree. To each village chief I offer two red and one white nut, wrapped in a piece of leaf.

Cola nuts symbolize respect, and they are also a powerful stimulant for those who chew them. Their symbolism is the main thing: I am showing respect to an Elder. After receiving this traditional honourable symbol, the Elders will be ready to listen to new ideas.

On a visit to Tunis, my wife Jeanne went up to a policeman to ask the way to the bank. "Alors, on ne dit pas Bonjour?" was the reply. Jeanne was mortified: "Of course you are perfectly right," she stammered. "One should always say Bonjour, I am terribly sorry." The policeman was quite cheerful after that. Apparently the Westernized city of Tunis has not lost the capacity for good manners.

When my African friend Lamin went to New York, the boot was on the other foot. Arriving at Kennedy airport, he went up to a policeman to ask the way to the Limo station (limo = limousine = taxi). "Good morning, how are you today?" asked Lamin, as befits any well brought up West African.

The policeman looked down at him, stopped chewing, and snarled: "What's that to you, buddy? My day is my business!" An introduction to the Western way of life. Welcome to America!

ALBINO TWINS UNDER THE SUN

Twins are popular in the Sahel. If you give charity to twins, people say you receive double blessings. Down in the dark mysterious forest lands, twins are considered a menace. Presumably this is because a woman who tries to feed two babies, may lose both if she has not enough milk. In the effort to feed two infants, she may use up her own reserves of health and strength and put the whole family at risk. Twins in Cameroon used to be left out in the forest for lions or jackals or hyenas.

This practice went out fashion when the White Men came and imposed a new way of thinking. Now don't get me wrong: I am not in favour of putting twins in clay jars and abandoning them in the bush. But Western thinking is often rather dry and partial. David Werner remarks laconically in his wonderful book "Where there is no Doctor" (Macmillan and Hesperian) that, "Twins are often born small and need special care. However, there is no truth in beliefs that twins have strange or magic powers." That is a medical opinion. Who can say in any particular culture what is a magical blessing for a family, or what may be dangerous to the well-being (mental or physical or economic) or solidity of a family structure?

In the center of Bamako city there is a vast space known as Square Patrice Lumumba. I tested this out on my 14-year-old Chet and some of his African classmates, and was impressed that they knew that Patrice Lumumba was an assassinated Congo-Zaire independence leader. So, their schooling is not exclusively French grammar and dictation, although they were not told that the assassination was set up by the American CIA with Belgian soldiers. Square Lumumba is the home of two albino boys. Twins. Well maybe "home" is a not the correct word, because their parental house is in a suburb on the other side of the river, but Lumumba is where you will find these boys from 7 in the morning until nightfall. They are especially visible because of their skin color, known here as "blanc local" - as opposed to a "blanc étranger" like me.

When they were babies, the boys used to sit under the shade of a magnificent colonial avenue of silk cotton trees – or kapok tree (*Ceiba pentandra*) similar to the famous 400-year-old Cotton Tree in the centre of Freetown, Sierra Leone. The trees of Square Lumumba were the glory of Bamako. In 1990 the French company Satom increased its annual profits by bulldozing all 60 trees out of Square Lumumba and covering the area with laterite and tarmac. For decoration, Satom surrounded the square with concrete blocks at one-meter intervals. Previously Lumumba was a place to park your car in the shade in front of the French Embassy. Hardly anyone parks there now. We avoid this sunbaked parking lot, where the car becomes like an oven after five minutes. After ten minutes, you cannot grasp the steering wheel. Breathing becomes an effort. The cars's rubber tyres stick to the soft tarmac. Women's thighs are scorched by the heat in the car seats. Satom could just as easily have made their tarmac parking lot under the trees, but contractors make a lot of profit from bulldozing trees, so Satom insisted on destroying the glory of Bamako. Moussa Traoré's corrupt military administration complied with Satom's desire for profit, and selling the wood made a profit for some colonels. This is a vivid example of how Western companies mis-spend "development" money in partnership with local political elites.

My daughter Leïla was so incensed about the destruction of century-old trees, that she wanted to write a protest letter to Mali's President. I said I had no objection. It might get me thrown out of the country: that has happened before (although not to me). In general, African Presidents are more mature than they were in the early days post-independence when criticism was unacceptable from anyone, even from a brother. I thought President Traoré might invite his ten-year-old correspondent up to the palace for a "photo-opportunity" and make a declaration about the importance of planting new trees. Writing letters does not come easily to Leïla. As is the way with ten-year-olds,

Leïla kept saying she was going to do it.....she hadn't forgotten.....she was about to start the letter.....until Moussa Traoré was overthrown by the people of Mali in a popular revolution. Leïla is now wondering whether she should write to the new interim government. I said I had no objection.

To return to our albino twins, their "home patch" is now sunbaked desolation where black tarmac soaks up the heat of the sun and bakes them. Their patchy whiteish skins and weak eyes suffer painfully from the Sahelian sun and heat, but this black tarmac desert is where they make their livelihood. The twins make a respectable living from charity. Whatever Western medical opinion may say about twins, they are believed to bring luck around these parts. Especially albinos. Albinos have special powers. It is not entirely clear to me what powers they have: Malians are rather vague about magical powers in general, although they are very clear that such powers do exist.

A hundred years ago, a King of Ségou called Monzon Diarra kept albinos in a pit under his throne room, because of their magical powers. Whenever he banged his staff on the roof of their dungeon, the albinos groaned aloud, giving the impression that King Monzon Diarra commanded the spirits of the Earth. Unfortunately for the albinos, the King of Ségou believed the way to extract their powers was to burn them alive. Such practices have died out. The albino twins of Square Lumumba may be in danger of roasting alive on the burning tarmac, but they will not put to death. They have been there since they were small children and they're about 15 now. They used to be very shabby. Last year somebody gave the boys identical smart, bright green tracksuits and matching baseball caps, so now they are visible from one end of Square Lumumba to the other.

My friend Konimba often gives them charity. "I am born after twins of my mother. So, if I see some twins, then I have to give them some charity. Maybe one time I will buy two chickens if I have enough funding. Then I will go to the twins and cross my arms and give one chicken to each twin. For twins you always give two of something, the same to each twin."

Every morning in front of my office, there is a mother who sits with her 8-month-old twins and two small plastic dishes. This morning one twin was lying asleep on a cloth beside the mother, while the other was suckling. At 7 o'clock in the morning there are already plenty of flies, especially because the rainy season is also the mango season, with masses of fly-food available. The sleeping twin had a couple of flies feasting at his nostrils. The suckling infant had flies around his eyes and one at the corner of his pursed mouth, trying to get into the milk business. I greeted the mother in ritual mumbling, "Is your health good? Was the night peaceful?" I bent down to the twin plastic dishes, crossed my wrists and dropped a coin into each.

A female colleague saw my gesture as she crossed the road, wearing magnificent red and blue robes and an orange turban wrapped around her plaited hairdo. "Zees is verrry good, Roberts! You weel get a double blessing. And if you continue to give charity to ze twins, then you will also have twins." Am I tempting the fates with my charity for twins? For some people this fate might be a blessing, but for others the idea of twins sounds like a menace. My wife Jeanne is horrified. I have been ordered to stop tempting the fates forthwith.

KEEP YOUR MAGIC HAIR ON!

I am sitting in the barber's shop in Bamako having my hair cut, day-dreaming about ripe mangoes hanging from massive green-shrouded branches above my hammock, where the spirits of many past ancestors must have rested on their way to a better place (spirits are believed to rest on the branches, not in my hammock – although the male ancestors probably had their own hammocks to lie in when they were alive). My mind is lulled by the snip-snip of scissors. I am rudely awakened from my reverie by the barber shouting at a young man to "put down that hair at once." In the mirror I see the embarrassed victim of this harassment remove from his pocket a lock of my hair, and drop it back on the floor from where he had surreptitiously picked it up. As it happens, the young man is my driver Doulaye, who is waiting for my haircut to end. Why would Doulaye want a pocket full of my hair?

"You cannot be too careful," explains the barber as his snip-snipping continued. "I always keep a close watch on customers' hair. Every evening, I collect it all up myself and burn it personally so that it cannot fall into the hands of anyone wanting to play tricks or do magic." The little Frenchman has been in Africa for so long, that he remembers shaving the arrogant chins of colonial Governors. He knows Africa. He proudly describes ancient invitations to the governor's mansion to celebrate France's national day on July 14th. Although I do not share his colonial attitudes or his French cynicism, I enjoy probing him to hear his stories and his prejudices. I've also had enough magic done at me and around me in West Africa, to know that he is serious about the business of burning hair.

Only last month, I went with Old Brother to the funeral of a chef de quartier in Magnanbougou village. "He wasn't very old," I commented sagely. I know how to set Old Brother talking. Sure enough, he responded: "No, I think he was under 45 years old. They say he was carried off by a flèche indigène." Literally translated, this means 'an indigenous arrow' and it is a euphemism for poisoning or magic. In other words, Old Brother suggests that a political murder was carried out using local methods. The police leave these cases alone, and I don't blame them. I keep an open mind. Maybe the village leader simply died of natural causes? Quite a lot of West African men die in their 40s from liver cancer, caused by hepatitis C or by poisonous aflatoxins growing on mouldy, poorly-stored groundnuts.

Africa has lots of magic, but it is far from an African exclusive. Locks of hair have been prized in Western Europe for thousands of years. A lock of hair can help cement love between two people (a mother and her baby, or a pair of European lovers). Some believe that a lock of hair will bewitch an unwilling or hesitant suitor, and bring him or her around to seal the deal. That of course is magical thinking. Harmless magic? Well, it can be. It may not be so harmless if you are trying to steal somebody else's lover.... and there is a lot of that going on in Mali. Love potions provide the largest slice of the 'magic market' in Bamako. Economists would be astonished by the cumulative total value of the love potion market in this city.

Thinking of my own cultural heritage of magic and mystique, I thought I had better seek advice about hair collection. In short, what did the young man Doulaye want with a lock of my hair? He said he had been asked by the tailor to collect some of my hair, so that the tailor could become rich like a White man. That seemed like a plausible story, but I wanted to learn more.

I walked into the village with a bottle of cold beer and two glasses, to visit a knowledgeable neighbour. Mr. Niakaté is a specialist on plants and recipes and there's nothing like a bottle of beer to get him talking. Well, there is, actually: he prefers whisky! I opened my bottle of beer, and we sat outside Niakaté's house in the cool of the evening beneath a convenient street lamp buzzing with moths. After a few minutes of inconsequential chitchat, I asked about the value of hair.

"Der is no problem with hair. It is used for making people strong. For example, eef we take the hair of a White man, then we can gain his intelligence. The marabout takes the hair and he burns it and puts the ash into some water. Then he writes Islamic words with the liquid on a board and washes it off again into the bowl. Then eef you use this, you may get rich like the White man or become very clever like him."

Niakaté grinned delightedly at the thought of becoming rich like me. His teeth gleamed yellow from the incessant chewing of cola nuts, his gums looking dried and gritty, putrid with their stain. I thought of my Old Brother next door, who is a truly wealthy African and the centre of a rich network of influential relatives, ministers and businessmen. When Old Brother travels to Paris, he flies First Class. I compared him to my own modest charitable salary, and I laughed privately at Niakaté's distorted stereotypes about wealth.

"And then you drink this hair mixture to make you strong or rich?"

Niakaté's grin faded. He looked pityingly at my ignorant head and explained patiently, "No, you do not drink it. You wash yourself with the liquid and the strength goes into you." Well, that's one good thing about the recipe: at least the barber's pilferer (my driver) wasn't intending to mix a cocktail with the unwanted scraps of my unwashed hair picked off the floor.

"Sometimes people will pay a lot for the hair of White men," continued Niakaté with satisfaction. What with his numerous children, his chewing of cola nuts, his incessant smoking of cigarettes and his love of beer, Niakaté is always broke. He talks a lot about money for this reason. He talks about alcohol even more. "You may offer me anything to drink," he once explained to me, picking up a bottle of Pastis and caressing it lovingly: "But you must never, never offer me anything weaker than beer."

On occasions such as this, I can be a pretty quick thinker. I offered Niakaté a Pastis, which he accepted with alacrity. "No, he cried, don't bother to put water into it!" Pastis is normally diluted 1:3 or 1:5......but Niakaté drinks it neat. I wondered vaguely whether Niakaté had ever thought about cutting off a piece of my hair in the hope of getting rich enough to keep himself in Pastis. Perhaps Doulaye the driver was collecting my hair to sell to Mr. Niakaté?

"Marabouts who live in the bush far away from the capital find it difficult to obtain White men's hair," said Niakaté. "If they need some urgently for a reason, then they may be willing to pay even 2000 Fcfa for some toobab hair." That would be something like $8 for a handful of hair: two or three day's wages for an ordinary Malian (at the rate of exchange in 1990).

The word 'toobab' – from the French slang word toubib, meaning a doctor (originally from an Arabic word) - has become the general term in West Africa for 'Whites' (foreign White people, not local albinos). The market for selling toobab hair to rural marabouts must be profitable: at $8 a tuft, someone could make a small fortune just from my hair droppings. Come to think of it, someone – my driver - was probably trying to do exactly that, until the barber stopped him. I am glad the barber caught him: black magic can also be done using hair. If your hair falls into the wrong hands, nasty things could happen.

Magic brings a good living to Malians who are good at it. Nowadays the main practitioners are marabouts, Muslim purveyors of charms whom my Malian friends often translate as "charlatans." They mix traditional beliefs and Arabic writings to provide treatments that can be harmful rather than helpful. The sale of love potions and love amulets is popular and profitable. The plausible man with the rosary from Mecca makes a lot of money, but he is not a genuine practitioner like the bush healer.

In the African countryside, those who know their healing plants are Hunters, Blacksmiths and other senior members of rural Initiation Societies – women as well as men. They practice their craft as healers, herbal pharmacists, intuitive psychiatrists, and marriage consultants. Blacksmiths are also famous as magicians because they are linked to fire: by combining air with fire, blacksmiths perform the magic of transforming earth into iron. Blacksmiths make indispensable items like hoes and plowshares, horseshoes, bridle bits and stirrups, and weapons like swords and spears (and also firearms) …. and some are also goldsmiths who create jewelry. Blacksmiths are feared and admired in West Africa.

Blacksmiths also carry out ritual circumcisions - female as well as male blacksmiths. They play an important spiritual role in traditional Malian society, influencing (even controlling) important human relationships with the Ancestors. They may also impose social discipline, for no one dares to lie to a blacksmith: they know too much. Sometimes those who practice healing, also use poisons or create magic that does harm.

Modern doctors are at last beginning to take seriously their scientific bush brothers. Effective African herbal remedies are receiving more recognition than before. The Faculty of Traditional Medicine at the University of Bamako produces a wide range of herbal products, and I have used several of them. Maybe new university partnerships will bring forth a genuine African contribution to medical science, and some measure of independence from the usual exploitation of African resources by Western pharmaceutical companies.

People ask me sometimes whether I believe in magic. Like many things concerning "belief," I am sure of only one thing: I'd be very stupid to dismiss something simply because I do not understand it. African magic isn't a part of my European culture. African culture runs deep. Anybody who wants to work in rural development needs to seek understanding of the societies in which we work. Everything is different. The climate is hot and tiring: you had better not try to work in Timbuktu like in Stockholm, or you'll fall sick. Women are not treated the same here as in Montreal. Whether this is good or bad (which is a matter of opinion), this is the way the village works and that is a matter of fact. A Malian Mother is considered to be like a Goddess because she gave you life. In Montreal and New York, women may have washing machines but no one treats them like a Goddess. Malian villagers live with a fear of magic, surrounded by superstition and dangerous spirits that threaten harm to their children. This fear adds complications to the lives or ordinary villagers. We had better know all this, and take account of it. We ignore it at our peril. If you dismiss these fears, you are making a terrible mistake: the next flèche indigène might just be aimed at you.

RITUAL SACRIFICE IN MARRIAGE

Deep in the forest lands of Ivory Coast, a bridal procession is moving towards the village square. The sun filters thinly through the tree canopy far above our heads. There is not much of this forest left these days: greedy Swedish lumber companies and venal African leaders have devastated these lands of the Baoule and the Koulango, not far from the legendary gold kingdom of the Ashante (Ghana). The procession of young girls sings a mournful bridal song:

"Oh my father, Oh my mother,
What harm have I done you, that you
should abandon me into the hands of
a stranger?"

From the village clearing the voices of the married women reply in unison:

"My daughter, you are not abandoned.
Every daughter is destined to leave her mother
that she may become a mother.
I left my mother to give you birth:
one day your daughter will leave you,
for that is a girl's destiny."

In front of the assembled members of the lineage, the bride kneels in front of her groom and lowers her shaven head. Then her father raises his voice: "I give you my daughter with a good heart. From this moment she is yours, to obey you, and to do your will. If she should fail, you can kill her. Now take her and do with her as you please."

And the groom replies: "I accept your daughter as my wife, and before all the community I promise to protect her and to support her and her family as long as I live." The patriarch hands to the groom the hair that has been shaved from the head of his bride. The groom raises his bride with his left hand, saying, "Rise up, my wife, and remember that you are now my property, and

I have over you the right of life and death." And the drums beat with renewed frenzy, the married women dance with new energy, and they shout and sing to celebrate the passage of a virgin to become a new member of their group. And the men drink.

I am happy to confirm that husbands no longer have the right of life or death over their wives. Marriage has changed: motorized travel, transistor radios and the power of television have taken urban influences into the remotest villages. But it is helpful to remember ancient marriage tradition when trying to understand the marital problems of your friends in West Africa's cities.

Sacko tells me (really he is complaining) that his wife "does not help me at all," which means that she has a salary which she spends on clothes, while he is expected to pay for rent and food and the costs of children when they come along. Many urban women expect their men to cover all the costs: when they are courting, they often set the standards by insisting, "You find me beautiful because I have a fine hairdo and jewelry: and if you want me to remain beautiful, you must pay my hairdresser and increase my stock of jewelry." It is expensive for a young man trying to find the love of his life in modern Africa.

Some of my friends have modernized – or urbanized – their marriages, but not many of them. Even among love matches, my friend Sekou is unusual when he says: "Yes, my wife helps me a lot and since she is an accountant, she even looks after the money we make from our grain mill."

I couldn't say that Jeanne "helps me" at all: our marriage is a partnership where we both share costs because we have a single joint budget. My friend Sekou is the only African I know whose marriage works like that. All my other city friends in West Africa have a Male budget and a Female budget. The Female budget covers hair-dressing and braiding, clothes, jewelry, perfume, incense and other intimate necessities. The Male budget covers all the living costs, food and housing, schooling and clothing, and transport and cinema money for the rest of the family (often including the wife's relations!).

Most people find it hard to make ends meet through the year. Marriage becomes especially stressful when there are special expenses like Tabaski: The Feast of the Sacrifice, or Id el-Qebir, which fell around June 11th this year (according to each sect's interpretation of the Moon's schedule). The religious sacrifice involves the slaughter of a male sheep to honour God, and to distribute meat and share meals with poorer neighbours: but Tabaski is also a time of sacrifice for husbands.

Tabaski commemorates Abraham not killing his son Isaac (according to the Koranic version of this story, it was actually his other son Ishmael – but why should that detail bother anyone?). At the last moment, just before killing his son, Abraham spotted a ram conveniently caught in the bushes by his horns. Abraham sacrificed the ram instead. In celebration of that original Old Testament ram, across the world some 40 million of his successors are slaughtered every year to celebrate Tabaski. Every married man (or householder) is expected to kill one. Many kill a goat because they cannot afford a sheep - a humiliating admission of economic failure. A good goat is just as good eating as a sheep, but it doesn't have the same religious value: in social terms of course, not spiritual. Like Christmas in the West, Consumer Tabaski has overtaken Religious Tabaski in the cities of Islam.

Living at the margin is painful in Bamako. Often salaries have not been paid for several weeks and usually they are inadequate. While successful husbands have two or more sources of income (like a salary and some rental property, or a grain mill), many urban survivors have no work at all. Villagers drift into the city, settling into slums with brothers who already barely feed their children. Occasional labouring jobs help the joint male family income. But even a small sheep in the capital city before Tabaski costs more than a primary teacher's monthly salary: no occasional labourer can afford one.

This is where marriages come under stress. Urban wives are not content only with a sheep. They want nice new clothes for themselves and a new suit for every child. They need to give gifts to their aging parents (and preferably to buy them a Tabaski sheep as well). Plenty of Malian men have more than one wife, and these wives demand money from their husbands to pay for the clothes and the gifts. Credits are requested from tailors, fabric merchants, butchers and rice traders. There are few urban husbands who can cover the costs of Tabaski without borrowing money.

Back in the village, extended families get together and kill one sheep for a joint feast. Even the widows in the village will receive a piece of meat from a neighbour. Not so in the city slums, where everybody wants to parade her new set of clothes (and this year, it is not just a dress and some shoes: plastic handbags from Taiwan are a MUST fashion item for small girls among the middle classes). Once upon a time, the husband in the village had the power of life or death over his wife. Now in the city, the wife nags her husband to death.

The sins of the fathers have come home to roost in the homes of urban Bamako: the city brides of today know well how to avenge their rural sisters, especially at Tabaski-time.

AIDS, THE TWENTIETH CENTURY BLACK DEATH

Condoms are really the only protection against AIDS, the great plague of Africa. One of East Africa's bestselling condoms is called "Rough Rider." It says it has knobs all over the outside to give a "hard sensation." Evidently the marketing men are not aiming at the health and AIDS protection market: they seem to be encouraging the less attractive side of male hormonal drives. To be fair, I should add that when I asked about this in Nairobi, people were quick to say that Kenyan ladies are good customers for "Rough Rider." How could I judge for myself? To find out, I thought I ought to buy some "Rough Rider" samples for my own personal research.

Standing discreetly in a corner of the Nairobi store, I studied the packaging with care. Where are they made? No country admits to manufacture. The Indian shop-holder denies knowledge of origin. Are they made in India? He doesn't know. But I want to know. Where do the condom-cowboys live? Where do they manufacture stuff with knobs on?

It is not just idle curiosity. "Love safely" says the advertisement. How does one love safely? Condoms are the only answer. But if the condoms are to work, they must work! The World Health Organisation has tested some of these condom varieties, and their research finds that condoms on the East African market are not up to standard. They are not strong enough. They cannot stand the heat of battle. Mid-summer in Tanzania was celebrated with the arrival of a $10,000 machine to test condoms. Your minds are boggling: how, I hear you ask, do they test a condom, if not....? Well, it is all a question of electric currents (or perhaps electrons) being blocked by the "sock." If they can pass through, then the "sock" is not water-proof (if you get my meaning).

The American development agency USAID is a major supplier of condoms in Africa. In most USAID offices there is a free supply for all employees and their friends. Just ask an American, and he'll probably be able to help. If you cut out the Indian cowboy sellers, you will save money and reduce risk (in USA they have lots of those $10,000 machines: if you use a "made in America," you can be pretty sure the electrons have been there before you). Like all bureaucracies, USAID makes mistakes. The BBC World Service told us with a touch of humour, that USAID during the 1990s imported to Tanzania a container load of condoms that were too small for local use. They had actually been made for the market in South Korea, and mis-routed ..... (OMG)

For curious readers, I can reveal that here are four standard condom sizes: S (small, or "snug-fit" - a frequent Asian purchase); M or "Regular" (most Europeans are "Medium Regular" customers); L (which is the standard size in Africa); and XL condoms which are probably bought by porn stars and freaks. Each should seek his own size: but if you need a size L, you would be well advised to buy your personal stock before traveling in Asia.

American interest in AIDS seems to have originated in San Francisco. Many Africans truly believe that AIDS is a capitalist invention. Some believe that Americans sent AIDS to Africa to annoy the Africans, or to replace them (American racist neo-cons have written articles to suggest that Black Africans should be replaced by Brown Asians who, they think, would produce more food to feed America's Empire). Some West Africans still believe that AIDS doesn't exist, that AIDS is Western propaganda. Anglophone West Africans call it "Slim disease" (because it wastes you away). Francophones call it "Le SIDA." This is an anagram of AIDS. Unfortunately for Sweden, SIDA is also the logo of the Swedish International Development Agency.

"Le SIDA? Ça n'existe pas" a young man told me in Dakar, capital of Senegal: he doesn't believe in AIDS. No one in East

Africa thinks that. Although some statistics seem to place the USA ahead in terms of AIDS deaths, it is really in East and Central Africa that AIDS has been ravaging the population. The difference is partly in the statistics: in the USA, they count them; in Tanzania, they do not.

A medical friend of mine accompanied a UNICEF mission deep into the Kahama region of Tanzania several years ago, where the dry savannah lands spread along the Burundi borders from Lake Tanganyika to Lake Victoria. The UNICEF experts were looking at child and community health priorities. What they found was so shocking, that one of the doctors made a video: otherwise no one would believe him. "We only have one priority now," said an old man: "It is the survival of our village. All the men and women have been carried off by the sickness. Only small children are left, with us their grandparents. We are old. We cannot work the fields and raise a new generation for our community. If you wish to help us, then bring us adults to grow our food, and people who will care for the small children so that they will be able to grow, and to produce future generations for us. For if you do not do this, our community will vanish."

There was an AIDS absurdity in East Africa: Tanzania, Burundi, Rwanda and Uganda were fighting huge AIDS epidemics, while Kenya's President Moi maintained that his country had no AIDS. Why do African politicians make themselves look foolish? Anyway, Kenya has now woken up to the gravity of Africa's AIDS catastrophe. A study in Nairobi found 60% of prostitutes to be seropositive. In Rwanda 20% of pregnant women were HIV positive: an even more devastating statistic. Another report claimed that 80% of the Zimbabwean army were seropositive: who would doubt it? That could be even more devastating, given the reputation of the army. The hospitals of Zambia were crowded with AIDS patients. Zaire was reeling under the AIDS plague. One of Africa's bestselling music cassettes was the 15-minute track called Le SIDA, recorded by the famous Zairois singer Franco before he died of an "unidentified wasting sickness" in October 1989.

"Oh, Le SIDA, un mal qui ne guerit pas..." sang Franco. He was right to sing that AIDS had no cure. I listened to Franco one warm and sultry evening at the open-air village cinema, before watching a Hindi love movie (the choice was Hindi or Kung-fu, and I loath Kung-fu). A Rwandan girl-friend called it "Le mal du 20e siècle" (she could have been talking about Kung-fu, but she actually meant AIDS).

"But there is no doubt that AIDS has changed our sexual practices," she continued with a wry grin: "In the old days, if his wife was away, you would think nothing of hopping over the fence to visit the neighbour for a bit of fun. We do not have the same taboos about sex as you Whites. But now, that has changed. With *Le SIDA*, no one goes visiting like that anymore, unless they are crazy."

The population explosion has long been on top of the list of Africa's development problems. Until climate change began to hit the Sahel, most African countries were able to feed their populations unless the rains failed. If a nation cannot feed itself, it is usually because of incompetent government and poor distribution systems. But feeding the population is not enough. If the population is growing, food production must grow too. Kenya used to hold the world record for population growth, with an average 3.4% annual net increase. Now it is the countries of North Africa and the Sahel: all around the Sahara Desert, you find frightening numbers of young people looking for employment, dreaming of a future that they cannot see at home. So they emigrate. Many young people dream of creating a new life in Europe. While Europe's population explosion is behind us, and our non-working elderly are increasing in numbers, Africa's young population is blossoming. Africa is already supplying Europe with many of the workers it needs to keep the economy working. But long-term planning has to face up to this new element "*Le SIDA*."

The best and most influential journal in Africa is the French language weekly *Jeune Afrique*. Its influence can best be measured by the fact that it is banned or seized at least once each year by most French-speaking dictators in Africa. "Stand up the government which has not attacked *Jeune Afrique* in the last twelve months!" One year at the end of the 20th century, *Jeune Afrique* had a four-color cover feature on *Le SIDA*. On the front cover they printed the following: "In the next few years, 220 million Africans may die from *Le SIDA*. Yes, that is not a misprint. We do not mean 22 million: it really is 220 million."

The greatest influence on the history of medieval Europe was the Black Death, which wiped out 25% of Europe's population during the 14th Century (and 50% in some areas of central Europe). Developing Africa is standing face to face with a similar calamity: the combined threats of AIDS and climate change. UNICEF estimated there were already 10 million African AIDS orphans before the year 2000 arrived. Poor, malnourished and ill-educated, how will they survive as adults during the 21st century?

If they cannot live decently in Africa, many of these young Africans will migrate to Europe. The combination of impoverishment through colonial rule, the unfair rules of international trade, the increased poverty due to AIDS, and the impact of Western industrial life on climate change are making it impossible for many Africans to thrive where they live. African Migration is an existential threat to Western democracies, because decades of Western bad policies and neglect have produced an existential threat to life in Africa.

NOTE from the author in 2024
In the 21st century, there are treatments. Among thousands of internet references, I found this helpful update on 14th July 2024 on CNN https://edition.cnn.com/2024/07/24/health/lenacapavir-preventing-hiv-in-phase-3-study/index.html: "Two shots a year of a drug currently used to treat

HIV infections were dramatically effective at preventing infections in a study among young women and adolescent girls in Africa. The twice-yearly injection of the drug lenacapavir can provide total protection against HIV infections, demonstrating 100% efficacy in Phase 3 trial data released by drugmaker Gilead and published Wednesday in the New England Journal of Medicine."

People who are HIV sero-positive can now live for a long time with medication. Africa still suffers, of course: UNAIDS reports that in 2023, 64% of people living with HIV were in sub-Saharan Africa, and about 62% of all AIDS-related deaths from the disease occurred there. But there has been tremendous progress. The number of people in sub-Saharan Africa becoming newly infected with HIV went down from 2.1 million in 1993 to 640,000 within 30 years — a 70% drop.

DRACUNCULIASIS, A PLAGUE OF FIERY SERPENTS

Jean-Baptiste was moving faster than I was. We were walking along the fascinating Dogon Cliffs on our third day. Dogon country is a great place for a walking holiday, following dramatic cliffs along the Burkina-Mali border. I meandered between sculpted village houses, avoiding holy fetish places, negotiating with local blacksmiths for their wares. Meanwhile Jean-Baptiste marched tirelessly from one village water pump to the next. My surprise at his speed was not because he was carrying my bag (while I wasn't carrying anything heavier than a water bottle): this is his terrain after all and he is about thirty years younger than me. What amazed me was his ability to continue walking, while nursing a hideous running sore just above the ankle. Believe me if you will, but even a syphilitic sore cannot beat the Guinea Worm for hideousness. (Well, it's not worth arguing about. Let's drop the comparisons since ugliness is in the eye of the beholder, and especially the sufferer.)

Jean-Baptiste had caught the worm's head and attached it to a twig, which was tied around his ankle to a string. Slowly, slowly, each day one-half twist, he was drawing the worm out of his body. Moses taught the Israelites the same method for dealing with their "fiery serpents" and medical science hasn't come up with a better solution since the days of Moses. That's rather ironic when you think of the medical profession's logo. Have a look at it, for example, on documents of the World Health Organisation (WHO). The creature climbing up the medical spatula is neither a "wise serpent" nor a cunning macho reminder of the treachery of Eve: it is a nasty little Guinea Worm.

Guinea Worm is agony (or so I am told: this is one of the local specialties I really want to avoid in Africa, together with yellow fever, blackwater fever, and brucellosis). For affected villages, Guinea Worm is a plague. Physically disabling, it is also one of West Africa's most economically crippling diseases. The larvae live in a freshwater crustacean called cyclops, which is almost too small to see with the naked eye. If you drink infested water, the larvae will spend about a year developing inside your body, migrating from the stomach mostly into the legs. The worms usually break out through the skin in the rainy season when there are puddles and ponds into which they can discharge more larvae - which need to find more cyclops, so that the Guinea Worm life cycle can continue.

The rainy season is precisely the season of agricultural activity, when men and women trudge out into their muddy fields to till, sow, weed, and harvest their annual food supply. Guinea Worm infected sores are very painful; when a local outbreak occurs, many of the active population are left lying on their sleeping mats in Guinea Worm agony, while the rainy season passes them by. In Nigeria alone, there are estimated to be 650,000 sufferers per year. From Mauritania through Mali, Ghana, Burkina, Benin and Nigeria to Cameroon, there must be far more than a million farmers each year whose agricultural production is hit by dracunculiasis. Dogon country has plenty of victims. The infection is so traditional here that the Dogon language has created a whole vocabulary for Guinea Worm, each stage of the disease having a different name.

Jean-Baptiste just shrugged when I asked him if it was very painful. I asked him if he knew how he got it. "Yes," he said. He must have picked up the parasite in a pond during the previous harvest season. He was thirsty, so he had to drink. Did he know the pond was infested? Well, he knew that you do not get Guinea Worm if you drink out of clean wells, so he supposed he knew the pond might be infected. Why didn't he carry a goat-skin or a water bottle like me? Jean-Baptiste shrugged again. Did he know that you can avoid Guinea Worm by the simple process of filtering the water through a finely-woven cloth (to keep the cyclops out of your drinking water)? Jean-Baptiste was not sure whether he knew this or not. Maybe he had heard it. Maybe he had seen a few of the women filter their water. In the Catholic Mission, they used to filter their water when Jean-

Baptiste was a young boy - but now they have wells and pumps, so they don't bother with filtering any longer, he told me.

It sounds like Jean-Baptiste has most of the knowledge he needs. He probably isn't "typical" (who is?). For a start, Jean-Baptiste has worked in the Catholic Mission, and now he works with tourists. At least my conversation with Jean-Baptiste shows that one fairly ordinary Dogon man has good basic knowledge of the problem and how to avoid it. He is ahead of the coastal villagers on the Gulf of Guinea, many of whom still attribute Guinea Worm to sorcery (understandable); rain (superficially true since the worm appears in the rainy season); failure to respect traditions; or adultery (which may give you AIDS or syphilis, but which could only introduce the cyclops under the most improbable anatomical circumstances).

Given Jean-Baptiste's existing level of information, it should be easy to spread the word on stopping Guinea Worm. If the education program was stepped up, with posters in every tiny village church and mosque, with the radio reminding people constantly about clean water, Guinea Worm could probably be eliminated. After all, they've eliminated it in West Indies and in Latin America (where African slaves are said to have brought it in with them... although one wonders if they also brought the cyclops, without which the reproductive cycle cannot occur). In Soviet Central Asia the Russians managed to get rid of dracunculiasis before the Second World War; and the authorities removed in from Iran during the 1970s, thanks to the Shah's White Revolution. The main difference between West Africa and these other countries is that in Africa, we have become dependent on the fancies of the donor agencies.

If the Americans or the British decide to support African health education, they design a short "project." African health educators have become used to the short-term fancies of those who provide posters, vehicles, and per diems for a one year campaign. In Europe, health education is a permanent affair. Giant posters promoting louse control, for example, cling to the walls of British public libraries and schools, just as they did fifty years ago (I suppose this also shows how ineffective the hygiene campaign has been in Britain .... and how very persistent lice are). Donors in Africa seldom seem able to concentrate long enough on any single subject for their efforts actually to make a difference.

Nigeria and Mali in the 1990s were two of the countries that launched plans to eradicate Guinea Worm, in partnership with ex-President Jimmy Carter of USA, who has turned the elimination of this dreadful disease into a personal crusade. In the summer of 1990, Carter and President Ibrahim Babangida of Nigeria made a joint declaration of war against the parasite. Their war chest of $10 million is piffling beside the billions blown away on wars. They couldn't even afford one B-52 bomber – but even if they could, it would be useless against the cyclops. The Carter Foundation is a private sector agency, and therefore it is more likely to stay the course than any government donor agency.

The crusade had other backers too, including two with reasonable track records for long-distance endurance: the United Nations Development Program (UNDP) and the Dutch Government. Since 1988 the Dutch have been supporting a long-term regional project called *rusafiya* = in the Hausa language, "water-sanitation-health." This campaign against Guinea Worm includes public education, pond treatment to kill the cyclops, and the provision of new village water systems across four of Nigeria's states. During the 1990s, the Bill and Melinda Gates Foundation joined the crusade, and the British government quickly jumped on the bandwagon. Now at the dawn of the 21st century, the Carter Center claims that Guinea Worm is the second disease – after smallpox – that is on the verge of disappearing from the world's list of scourges. Only half a dozen African countries still have cases of dracunculiasis. Fewer than 5000 cases are said to exist. The Guinea Worm crusade shows what can be done if people work together over a period of twenty or thirty years. I said 'people' because wars on disease and poverty and exclusion cannot be won by governments, or donors, or Americans. Development and the eradication of diseases can succeed only if they are part of a war on disease and poverty that involves and mobilises the people themselves.

ENERGY-WATER FOR GOOD HEALTH

ONCE UPON A TIME there was a man called Mamadou who lived in a village in the bush with his two wives, Bintou and Awa. Bintou had one son called Papa. After twelve years of marriage, Bintou had never managed to have a second child, and this made her very sad. So Mamadou took a second wife to bear more children, sons who would fill his compound with noise, who would create grandchildren, and who would look after him and his wives in their old age, when it is no longer possible for a man to plow the soil for days on end.

Awa had recently given birth to her first baby, a son whom Mamadou had named Moussa. Papa was very pleased with his new brother. But Bintou was jealous of the little baby Moussa. Why could she not have a second baby? Why had this young girl come into her household to replace her in the favours and favouritism of her husband? It was true that Awa gave no real cause for offence. She was polite to the senior wife. She cleaned well and washed well. Awa was careful; she never wasted any food, and her mother had trained her to be a good cook. In fact, Bintou had known Awa since she was an infant, newly born into their village. She had never disliked Awa, or her mother. But then Bintou had never imagined that this little girl Awa would grow up to become her co-wife. Now here was this second son, the image of his father: a baby who had arrived in her house through another woman. The baby was a symbol of Awa's fertility and youth, a guarantee of Awa's rise in influence. If she produced more sons, Awa would probably become Mamadou's favourite wife and confidante.

Bintou looked at the baby in her arms. Instead of feeling love for this small child, she hated him. In the social values of village society, Moussa was Bintou's child as well as Awa's child, and Bintou had the main care of the baby when Awa was not suckling him. Yet Bintou felt aggressive, rather than protective. What if this Moussa should grow up to be bigger and stronger than her Papa? What if Awa should have three or four children, leaving Bintou humiliated as the poor producer of a single child? What if anything should happen to Papa.... then what would become of Bintou? Would Mamadou not throw her out as a useless wife and a burden on the household? If the family had only Awa's children alive, what role would be left for an elderly barren woman like Bintou?

One day when Papa returned from the village school, he saw his mother alone in the compound. She was doing something strange. At the entrance to her co-wife's hut, Bintou was spreading a fine powder. Papa kept out of sight and saw his mother disappear furtively in the direction of a neighbour's house. There was no one else in the compound. What had his mother been doing? Papa was concerned. Instinctively he felt that his mother had been up to no good.

Two days later, the infant Moussa fell ill with diarrhoea. Awa was worried. Bintou brought a new amulet to hang around the infant's neck and said lots of comforting things.... but Papa could feel in his bones, that his mother was happy that Moussa had become sick. Had his mother Bintou done some evil magic? What could he, Papa, do to put things right?

The next day Moussa was weaker, the diarrhoea had not stopped. But the ways of God are indeed mysterious and wonderful, for that very morning there was a special health and hygiene lesson at school: the State Nurse arrived to talk about diarrhoea. Papa listened carefully, determined to bring help to his small brother lying at home. The State Nurse was a man with greying hair, which reassured Papa who had been brought up to respect the wisdom of his elders. Whatever this Old Man State Nurse told him to do, that is what Papa would do for Moussa.

"In the case of the running belly, said the State Nurse, the problem is to take water. The sick person is losing water. Therefore, you must give him water to drink constantly. Every five minutes, you must give him some water to drink." The State Nurse wrote carefully on the blackboard the numbers 1 to 21 and circled the 1 - 6 - 11 - 16 - 21. The children had to count out the numbers, so that they could understand that this meant "drink every five minutes."

"But water is not enough by itself." The State Nurse smiled: "if you are going out into the fields to weed between the rows of maize, will water be enough for you to eat? No! You need water, but you also need food to give you energy. And a sick person needs energy, too. A person with running belly cannot take food. It is too heavy and will make him vomit. So, we give him "energy-water" instead of food. If the sick person does not drink enough, they will die because they have no more water in their body, and no more energy. Now I am going to teach you how to make "energy-water."

The State Nurse opened a tin decorated with children wearing red caps and drinking milk. Inside the tin he had some special somethings, which he placed on the teacher's table. "What is this?" He held up a white square stone, which Papa had not seen before. "That is sugar," shouted Modou triumphantly. Modou had lived in the city until his father had sent him back to the village, to his grandfather's house.

The State Nurse explained how one small stone of sugar was the same as one small spoon of the powder sugar they could buy in the Cooperative Shop. He showed how you can mix the sugar in water, until the water drinks up the sugar. Then the children carried out the same experiment with salt. They saw that salt disappears more quickly into the water while the sugar has to be beaten with a spoon before it allows the water to drink it. "And even better than sugar, you can use honey. Honey is better than sugar because the bees have made the sugar easier to dissolve: in honey you find a special sugar called glucose, which is especially strong and good for people who are sick." Papa was pleased to hear this, because his big friend Amadou was a honey-hunter. He had no idea how he was going to buy sugar without getting into trouble, but honey he could get from Amadou, no problem.

The State Nurse was talking about boiling water. Papa knew from the previous health lesson that the bubbles show that the "microbes" are dying in the boiling water. Sick people don't need more microbes, that was obvious. Papa was getting impatient to finish the recipe so that he could go and try it out at home. At last, the State Nurse was getting to the final point.

"Now for a person who is sick with the running belly this is what you have to do. You take eight spoons of sugar or honey and two spoons of salt and you put them into a one-litre bottle of water. Remember: you must use boiled water, and a boiled bottle, to kill off all the microbes. Then you shake the mixture. How can you measure the salt and sugar if you have no spoon?" The children waited expectantly. "If you have no spoon, you use the cap of a beer bottle."

One boy asked a question: "What are the magic words to say when you mix the energy-water? Can only women do this, or the members of one particular age group?" To the children's surprise, the State Nurse said there were no magic words.

"The magic of energy-water is not in the saying, but in the doing. The magic of energy-water is in the bubbles which kill the microbes, and in the correct mixing of the honey and the salt. It is not blacksmiths or juju-men who do this best: it is people like nurses who know how to kill microbes. To succeed with any medicine, you have to be initiated. State Nurses go to nursing school to be initiated. It takes several years. You school students have been initiated by me today."

The students sat there spell-bound. None of the boys was yet circumcised. Here was a State Nurse giving them the first part of their initiation. They had not realized quite how important this energy-water was going to be. The nurse continued carefully.

"The best people to mix the energy-water and to give it are three: the mother of the sick person, or the big sister or brother if they have been initiated by a State Nurse. Energy-water does not work well with grandmothers. Grandmothers and aunts are good for certain medicines, brothers and sisters are good for other medicines. Initiation for old people did not include energy-water because it did not exist when they were initiated. This is a modern technique for modern initiates. This energy-water is best for the big sisters and brothers to give, or for the mothers."

The nurse wrote an arithmetic sum on the blackboard:

    8 caps of sugar + 2 caps of salt + 1 litre of water (boiled)

        = energy-water for running belly

Papa ran off to see Amadou before he went home, and got some honey. He also borrowed Amadou's watch. He collected some wood on the way home, so the women would not scream at him when he wanted them to boil some water. Then he went to see Bintou. He knew from what he had seen her doing in front of Awa's house, that she had something to do with the diarrhoea.

"Mother, I have been initiated into a medicine to protect Moussa against dying from his running belly."

"My son, leave God's work to God, and the ancestors' work to the ancestors. As long as you are two sons in the compound, your father's love will be divided, your father's wealth will be divided, your father himself will be divided between his two sons, just as he is divided between his two wives."

Firmly Pap replied: "Mother, Moussa is my brother, and Moussa is your son, as well as our father's son and our mother Awa's son. If it is in your power to stop the sickness, then you must stop it. If you will not help Moussa, then I will help him. I have been initiated, and I am going to try to save the son of Awa."

Papa's mother lit the fire and brought the bottle and salt he requested. She watched as Papa killed the microbes and manufactured his energy-water. As she watched, Bintou began to regret her own hatred of Moussa. If her own son wished to save the child, why did she want to destroy him? Papa was sitting beside his small brother, clutching Amadou's watch, and getting Moussa to drink every 5 minutes. Bintou went off discreetly to see an old woman about magic to counter the bad medicine she had made.

Little by little Moussa looked better. Thanks to the Sugar-Salt-Solution, the parched loose skin returned to normal, the sunken eyes became less noticeable, the child's lips and eyeballs began to look less dry. On the second day, Moussa was clearly looking much better thanks to his brother's energy-water. Papa was really beginning to think he had won the battle.

Then his father called him. "Moussa is looking better, and that is good. Awa tells me you are looking after him, and that is good. She tells me you have been initiated at school by the State Nurse, and you have learned a new medicine for running belly. That is good. At least school has been useful for one thing. Your grandmother did not want you to go to school, but I decided one person in the family should learn about the new things. You are doing well, my Papa."

Papa wasn't sure whether he was more proud about Moussa's recovery or about the extraordinary praise he had heard from the lips of his father.

TRYING TO FIND A YELLOW FEVER INJECTION

It was time for another Yellow Fever shot. This is not "yellow fever" in the parlance of the African Coast, meaning jaundice from hepatitis. Hepatitis is far more common than Yellow Fever. Hepatitis kills plenty of people, and it may very well account for the high rate of liver cancer in West Africa, which carries off so many men in their forties. Hepatitis is pretty dreadful—even pretty deadful. Hepatitis is not pretty, but it is nothing like the real Yellow Fever, which gets to be written with capital letters, which is lethal and which requires a vaccination that is renewed every ten years.

I remember how the last time we wanted a Yellow Fever shot, it was a hassle. We were in France when we realised that we were due for a booster innoculation. A few telephone calls told us we could get it between 9 and 10 a.m. on Wednesday in a city some 250 miles away from where we were staying. So, we were up at 4 a.m., kids and all (aged 4 and 1 in those days); we drove 250 miles; had hot chocolate and a croissant in café beside a gentleman whose breakfast was black coffee and a cognac; and found ourselves queuing before 9 a.m. in an unfriendly French hospital.

I said "Bonjour, Madame" to the receptionist. She looked up reluctantly, and glared "Sit over there." So, we sat. From 8.45 a.m. until 9.50 we sat with our two restless children, choking on the oppressive fumes of hospital disinfectant. Then we were injected by a silent male intern, and passed on to an unsmiling woman with glasses who snapped that our vaccination cards were invalid because they were printed in Persian. Well, they were perfectly acceptable in Persia. There was a translation into English, which she resented. So, we had to pay for new cards printed in France. We had to pay a huge amount for the vaccine and even more for the act of injecting them. We finally staggered out of the hospital at 11.45 exhausted, wishing we were back in smiling West Africa and not faced with another

250 miles to drive back, with a necessary stop in a restaurant for lunch.

Living in West Africa also has its hassles, but at least it is always friendly. I thought Mali might be easier than France, when it came to searching for a Yellow Fever inoculation. First, I had to find the Service des Grandes Endemies. Under this forbidding title, redolent with overtones of quarantines, Smallpox epidemics and blackwater fever evacuations, I discovered a dusty compound situated between the office of the United States Agency for International Development, and a vast smoldering rubbish tip. I ducked below the foul smoke and spied a down-and-out (a homeless person) asleep on a pile of rotting foliage, surrounded by a pile of empty coca-cola cans. His skin was grey with dirt, his hair matted, his shorts torn. Here was one of those who is classified in Africa as "mad" or "inhabited by a spirit," perhaps by a *jinn*. In old English such people used to be described as "touched..." (by a spirit, I suppose, or by God, or perhaps by the Devil Himself). The English easily dismissed such a person as "The Village Idiot." Westerners lock such people up in an institution—in London, it used to be "in Bedlam." In African villages they mostly leave the "touched" to wander around by themselves, and only chain them to a tree if they start giving trouble. In the city, these lost souls move around as the spirit takes them.

Holding my breath against the stench of the smoke, which covered any stench of the tramp, I dropped a coin onto the sleeping man's shorts and entered the Grandes Endemies. Two men were drinking tea: the sweet green tea of the Sahara Desert had been stewed over charcoal. They immediately offered me a glass. I sat beside them on the bench and slurped tea in the approved noisy manner. One of the men had a face pock-marked from smallpox fever. He had to be above forty years old, born before the greatest achievement of the United Nations World Health Organization, which was the elimination of smallpox. Poxy grinned at me with blackened teeth. I asked about Yellow Fever. "No problem. Every day from 8 until 10 a.m. See you tomorrow."

Next morning I went for my Yellow Fever inoculation, ready for a long wait: passport, health card, money, a new copy of the Guardian Weekly to read .... and my own new sterile syringe. In these days of You Know What, the prudent travel with a personal supply of syringes and condoms. I arrived at 8.02 a.m. The tea-drinkers hailed me as an old friend. "Hallo, Msieu, venez ici!" This morning, they were wearing white coats, installed on the wide veranda of the colonial building labelled "Grandes Endemies" sitting opposite each other across a square wooden table.

"C'est pour la Fièvre Jaune?" The one with the smallpox scars unzipped a new syringe from a box of 500, labelled "World Health Organisation." He plunged the needle into a vial, and turned towards me. He was a lot quicker than I was. I rolled up my sleeve, he wiped me with cotton wool, jabbed me, and it was done. The syringe was thrown into a cardboard trashcan for incineration. I was dismissed.

At the next table, a grey-haired man in a dark safari-suit was waving a ball-point pen. He took my vaccination card. We shook hands and established that we were born in the same year. I said we must be twins! He looked at my white skin, laid his black arm next to mine, and we roared with laughter. We shook hands again. Three rubber stamps thumped onto the card. The ball-point descended for a signature. "Voilà, Msieu." It was finished at 8.08 a.m.

I asked if I could pay something? "Non Msieu: le vaccin est gratuit." The vaccine was free. The syringe was free. The signature was free. Everything was free. Life is always agreeable in West Africa. It is always more smiling than in Europe or in USA. It is usually cheaper. Often it is simply a lot more efficient.

AFRICAN SOCIAL EFFICIENCY IS LINKED TO FRIENDSHIP

When we finally arrived at London's Heathrow Airport for our English Christmas, our luggage had not followed us. Flight delays and connections and baggage handling—the pressures were too intense for the Sabena system to cope. So, we had left the tropics for Northern Europe and we had no winter clothes. The children each had on an extra t-shirt and a light sweater decorated with holes made by the cockroaches. Jeanne and I doubted whether that would be enough covering for the whole holiday.

Heathrow was 30 degrees colder than the Sahel. Our children had never before worn a vest under their shirt, let alone gloves and scarves and bonnets and umbrellas. How were we going to survive the rigours of the English winter climate without our spare woollens?

We need not have worried. We reached Uncle Teddie's at 2.00 in the afternoon, and the telephone rang at 6 o'clock: "Hallo, this is Heathrow Airport. Your suitcases have arrived from Brussels and they will be delivered before 5 a.m. by Pony Express." Sure enough, when we opened the rain-sodden garden shed after breakfast next morning, there were the suitcases. What wonderful efficiency!

But I can match this example of European efficiency with its African equivalent. On our return to Bamako, we were met at the airport by my Old Brother. He is a big personality, who was an airline pilot in his youth. In the airport he wields considerable weight, and he carries an appropriately voluminous paunch under his flowing embroidered robes. Much embracing took place. How were the uncles, aunts, cousins, brothers, sisters, small mother (father's sister) and Uncle Tom Cobbly? They were all fine. And where were our passports? These were whisked away by a compliant immigration official, as we hustled through to search for our luggage with the assistance of friendly customs men.

So here we are, back in the human warmth of West Africa. Leïla is talking non-stop with her "sister" and best friend Selima, while the boys are exploring the piles of baggage. Old Brother is giving instructions in all directions and a few porters are waiting hopefully for money from me, the White younger brother, who will tip more generously than Old Brother. Everyone is smiling cheerfully and the customs officials are saying "Welcome." All is warm and friendly as only West Africa can be.

One week later, Old Brother ambled round in his embroidered robes while we were eating dinner. "My good friend Roberts," he beamed, "I have a surprise for you. What is here in my pocket?" I have no idea what might be in his pocket. A telephone bill? Yet another letter to translate? With the glee of a successful magician, Old Brother pulls from his embroidered chest pocket our five passports. My mouth falls open. "Yes, I too was surprised. Last week at the airport we all forgot to collect the passports. Then this morning one of the immigration officials dropped by the house on his motor bike and gave them to me."

I have tried to imagine a member of the British immigration service driving up to my house with the family passports, and I confess that my imagination has failed me. My memory of British immigration is dominated by a famous figure in the Consular Office in Paris, an absurd-but-sinister Fawlty Towers character boiling with anger. This aggressive man used to sit in the central glass cubicle, sporting a small military moustache, using the voice and oozing the charm of a drill sergeant. He spent six hours of every day shouting abuse at people who had the nerve to ask for a British visa. I found it painful to sit in the British Consular office holding my passport, listening to this man insult and abuse foreigners.

The French are equally as bad. Fatimata begged me to get her a visa for Paris. She is a busy African executive, and she cannot afford five hours waiting in the queue. "You will not have to stand for hours waiting in the Malian sunshine, Robert, because you are White. They will let you in immediately."

Next day I parked my car in the Square Patrice Lumumba, and looked at the line of forty Africans queuing outside the gate of the French Embassy. I walked up to the gate and asked: "Is the Consulate open yet?" An elegant Malian lady at the front of the queue replied: "It is open for you, but it is closed for us." I gulped, kept my mouth shut to hide my accent, and walked unchallenged past the uniformed gendarmes. The appearance of my White face was enough to set off a buzzer.... a steel grille opened....

The air-conditioned Consular Office was filled with friendly French females. Several recognised me as a parent in the same school as their children. One of them took my envelope immediately, and we chatted briefly about the up-coming school fête. She assured me that Fatimata's visa would be processed by the end of the week. All very efficient. Ten minutes after arriving, I left through the Embassy gate. The elegant lady at the head of the queue merely shrugged her shoulders.

Efficiency, I have decided, is selective. The French were efficient for me because I am in the right (White) group. They are inefficient for Africans seeking visas because they want to be (they want to discourage immigration). And so it is with African efficiency: it is highly selective.

Once she had her French visa, Fatimata traveled to the airport where she discovered for the first time that she needed a Yellow Fever vaccination certificate. She is a wonderful administrator and accountant, she is running a major rural development programme, but this was her first time travelling to Europe. She had never considered the problem of Yellow

Fever. Oh dear! What shall we do? An American or European would panic, cancel the trip, and seek an appointment for the vaccination. Fatimata had a more efficient solution.

On the other side of the customs office at Bamako airport, is the health office. Here they have health cards. For the paltry price of eight dollars, an immigration official was able to provide Fatimata with an appropriate Yellow Fever vaccination certificate. The objective was to catch the Air Afrique flight to Paris, and this objective was attained with plenty of time to spare. Fatimata's solution was most efficient. Even as I write, she is at her conference in London. I hope she does not have Yellow Fever. It is highly unlikely, since Yellow Fever is a tropical disease common in Africa's forest zones but seldom encountered in the hot, dry, savannah lands of Sahelian Mali.

There are those who, cynically, will comment that the miracle ingredient of Fatimata's success was not "efficiency" but the eight dollars. I agree that every engine requires oil, if it is to run smoothly. But Fatimata did not get her card because of the $8: she obtained her vaccination certificate because her fellow-Malian in the airport health office wanted to help her get to London. The immigration official who cycled over to Old Brother's house with our forgotten passports, didn't bring them for eight dollars. It was an act of neighbourliness. It was a social act, in a culture that gives higher value to human relations that to economic returns.

Efficiency is selective. Westerners usually think of efficiency in terms of cash. But is financial gain the only measure of success? The waste in Western big business is as huge as its profits (I used to work for a highly successful multinational corporation, where I saw plenty of waste). I think the chilly concept "economic efficiency" is a dubious goal. At most, it should be only one of many goals to be pursued by businesses. Perhaps the pursuit of friendliness is better.

In West Africa, it is social efficiency that counts for most. You'll not find anyone in West Africa who is inefficient about attending funerals, or weddings, or baptisms. There is no one who is inefficient about helping his neighbour in time of need. Never a week goes by without Fatimata helping a colleague with a bank-note to buy medicines for sick kids, or meat for the family sauce. Friendship relations are all important in West African society. Perhaps that is the key I have been searching for, to explain the difference between the European and the African immigration services. Both are efficient, as my examples have shown. But in rather different ways. In West Africa, efficiency is linked to a special social ingredient: Friendship.

CHRISTMAS DRIVING HOT AND COLD

I am driving on the M25 motorway around London in December 1992, on a dismal day of wind and sleet, surrounded by a family of malcontents. Jeanne's elegant nose and ears are glowing as she mutters: "All zees ees your idea, Robert." She makes the perfectly good name "Robert" sound like an African curse. "Fancy leaving Afreeca in ze mid-winter for zees gloomy cold! We should be lying bezide ze swimming pool. I 'ope you are pleesed wiz zees sleety alternatif?" Jeanne isn't feeling very warm or sunny. Is it my imagination, or does her charming French accent take on an exaggerated reptilian menace when she is annoyed?

The kids aren't happy either. They are bouncing up and down in the back seat of our hired car, chanting "We want snow! We want snow!" They've never seen snow. The only snow in West Africa is found inside the freezer; unless you include the artificial ski slope inside the ridiculously luxurious and hideously expensive Hotel Ivoire in Abidjan.

This M25 does great things to help travellers drive around London, but it terrifies me. In the African bush, I am mostly slithering around on sand dunes, keeping my speed up to 30 mph to avoid getting bogged down in warm soft sand. In Bamako town, normal driving is all 2nd gear stuff - in the middle of the road where possible, to avoid potholes, donkeys and goats. Here on the M25 motorway I am positively racing along at 60 mph. I am supposed to keep to the left. If I do so, external forces of traffic flow carry me inexorably off the motorway at the next exit and I have to go around a roundabout (African friends call it a "turn-table") to rejoin the motorway, and try again. So I keep to the middle lane, feeling like a hay-seed tossed around by the harvest wind, with vehicles racing past me on both sides at 70 or 80 or 90 miles per hour. Driving at night, the M25 seems to be composed of flashing orange laser beams that blind me and disorient me. We have no street lights where

I live in West Africa. Driving on the M25 feels like soaring through the middle of Star Wars. It is terrifying.

Such were my cold damp thoughts as I struggled through the speeding millions over the Christmas holiday, driving from auntie to pantomime to pub, to meet cousins and yet more cousins. But now I am back in Bamako, reinstalled in warm friendly West Africa, I can put the M25 horrors into perspective. Changing a wheel yesterday on the roadside, I realised that I had not had a single puncture during our three weeks in the UK. Crouched in the dry Sahelian dust and straining at the wheel-nuts, I felt quite affectionate towards the smooth wet tarmac of the M25. Here in Bamako I was struggling: with one of my wheels in a crater: the whole car was tipping sideways. I wedged the car wheels with rocks, balanced my jack on a flat stone to gain extra height, and started turning the lever again. Sweating as much with worry about the jack as from the burning African sun, I thought longingly of the cool weather and the extraordinary smooth surfaces of British roads.

Driving in Britain is a different ball game. Wing mirrors in Africa are used for shaving: they are mostly kept in the washroom at home. In Europe, wing mirrors are used for overtaking. In my West African town, the only overtakers are mopeds that weave their way between the potholes and the donkey carts. Fuel consumption in my hired UK Metro was derisory, less because it was small and new, than because I was cruising constantly at between 30 and 60 mph. In Africa most of my driving is at a juddering walking pace. And petrol in landlocked Mali costs 65% more than British unleaded at 40 pence per liter. There is no unleaded petrol in Africa. Most vehicles here are second-hand exports from Europe. Some of the overladen trucks on our African roads are 40 years old.

The hidden costs of bad roads in Africa are a major source of economic underdevelopment. I asked my cousin Chris in Europe: he never changes his shock absorbers. In Mali I have to change the shocks twice a year every year on my three-year-old

Peugeot. I also have to replace the rubber seals every ten weeks (mechanically-minded readers will be pleased to know that our shocks are filled with engine oil, which is better able to absorb the bumps, but which tends to break the seals due to its greater resistance). Because of the dust, the oil and fuel filters have to be changed several times each year. Because of the roads, I wear out 2 sets of tires per year. Sometimes an unseen pothole at 50 mph on the main road will rip a new tire from end to end.

The uneven road surfaces create a constant juddering that causes untold damage to door catches, window mechanisms, carburetor adjustments, and to the exhaust pipe. I reckon that I have the exhaust pipe refixed once every three months, and patched every six months because it has a hole pierced by grating a rock in the middle of some city "street." We patch it with flattened powdered milk tins, but it has to be replaced after in eighteen months when it has caught so many rocks that there is nothing left to patch. A new Peugeot exhaust unit here costs around 250 dollars (130 pounds sterling). After three years in the desert dust, I have just had to rebore the engine (another 1400 dollars). This is all expensive for me. Multiply it one hundred thousand times for all the other vehicles that are wearing out, and you have one hell of a lot of foreign exchange going out of Mali to buy parts. This creates a horrific picture of African transportation inefficiency in terms of lost time, lost goods, broken-down trucks, frustrated merchants, spoiled cargoes lying in the sun, etc.

And that is just the urban driving. In the sand dunes near Timbuktu, I burned out a Landrover clutch plate. Some years back, we hit a pothole on the road to Niamey, and broke the back axle. Once we got back on the road, a Fulani bullock took out one of our headlights with his curving horn. Well, cows are an acceptable natural hazard: bad roads, on the other hand, are the result of bad planning, inadequate technical skills, poorly-managed government services, plus bad Western development strategies. And sometimes it is straight corruption: like the EEC-funded road going north towards Timbuktu, built by an Italian company whose cemented fords were only one centimeter thick. They all washed away in the first rains, except for the first two, where the EEC Representative and the Minister of Transport had gone to cut the ribbon. I expect they each received a handsome commemorative gift from the Italian company. Senior European officials never discovered how their road became useless after the first rainy season, since they travel by private plane. But we bush workers use that road all the time, and it is hell. I'd rather drive through flat slithery sand than negotiate the jagged concrete edges around a two-feet-deep crater gouged out by the rains where there was supposed to be a ford. We pay coins to small children, who guide drivers around the deep craters, because the whole road is covered with water and rainy season drivers have no way of knowing what is road and what is a pond.

Our children were fascinated by winter in Europe. "Hey, Dad, there's lots of trees with no leaves on." These are summer children, who contrast the hot dry season of the parched Sahel with the gentle warmth of summer holidays in England, France or Canada, in summer landscapes clothed in green and bathed in sunlight. These children do not know cold; they swim ten months per year. Winter woollies are a new experience. Dusk at 4 p.m. was an exciting revelation. In the tropics, seasonal dusk variations move from 7.00 p.m. in winter to 7.30 p.m. in summer, and we call the seasons "wet" and "dry." Both are warm: 22 degrees centigrade at night in December, 37 degrees in June. Winter in Europe is a different challenge. There is nothing like winter sleet to remind children to pull on woollen bonnets, scarves, gloves, and gumboots. To escape the chilling North Sea gales, we snuggled down beside a North Sea gas fire, grilled crumpets, and remembered how hot we had been over a Malian Christmas when we hosted eight nationalities for lunch on Christmas Day, around a festive table for 19 people in shirtsleeves, devouring salads and turkeys. We sheltered from the sun under a straw roof, protecting ourselves from the heat behind a double screen of hanging bed sheets. Our Malian

Christmas trees were rich green mangoes, and erect papaya stems with fat fruits pendulous beneath a leafy umbrella.

Only the Sahel's acacia trees are as leafless as an English winter. Most acacia leaves have evolved into long thorns, which hold moisture from the sun and discourage grazing cattle. I discovered that the English holly tree uses a similar defensive system. "The holly bears a berry, as red as any blood..." I always thought it was a holly BUSH: but Uncle Teddie has a thirty-foot holly TREE in his garden, laden with red berries. Down at the bottom of the tree, the leaves are vicious: "The holly bears a prickle, as sharp as any thorn ...." Chet pushed his little sister into the holly tree to test out whether the carol song is right, and she yelled for twenty minutes. Once the leaves are above six feet high, where the cattle cannot get at them, they grow soft green and friendly.

And then there was the discovery of English food! In addition to the traditional Christmas meals, I discovered a new English iced cake. This cake was soft at the edges, and frozen in the middle because Aunt Zélie only took it out of the freezer at three o'clock, for tea at four o'clock. In Africa the freezer is an expensive luxury used to preserve food from rotting. In England it has become an expensive necessity used to spoil good cooking! Uncle Geoff has a micro-wave. We had never seen one before. Chet enjoyed putting cups of water into the micro-wave oven to watch them boil, while the cup stayed cool.

That micro-wave story shows you how far away West Africa is from *"Coronation Street"* - a television soap opera that has a micro-wave oven in every kitchen and which, amazingly, is still shown twice weekly on British television. More astonishingly, the British farming radio show *"The Archers"* was launched during World War Two to teach people wartime survival skills, and two generations later people are still listening to *"The Archers"* every day on BBC Radio 4. And Agatha Christie's play *"The Mousetrap"* is STILL running in London, would you believe, in its mindboggling 39th year! It seems there are some forms of tradition in Europe that are as timeless as the traditions of West Africa.

Made in the USA
Middletown, DE
18 September 2024

61137938R00144